T0269467

SpringerBriefs in Applied Sciences and Technology

More information about this series at http://www.springer.com/series/8884

Elías Cueto · David González
Icíar Alfaro

Proper Generalized Decompositions

An Introduction to Computer Implementation with Matlab

Springer

Elías Cueto
Aragon Institute of Engineering Research
University of Zaragoza
Zaragoza
Spain

Icíar Alfaro
Aragon Institute of Engineering Research
University of Zaragoza
Zaragoza
Spain

David González
Aragon Institute of Engineering Research
University of Zaragoza
Zaragoza
Spain

Additional material to this book can be downloaded from http://extras.springer.com.

ISSN 2191-530X ISSN 2191-5318 (electronic)
SpringerBriefs in Applied Sciences and Technology
ISBN 978-3-319-29993-8 ISBN 978-3-319-29994-5 (eBook)
DOI 10.1007/978-3-319-29994-5

Library of Congress Control Number: 2016933200

Printed on acid-free paper

This Springer imprint is published by Springer Nature
The registered company is Springer International Publishing AG Switzerland

To our families, for being so patient with us

Preface

This book is intended to make life easier to those interested in model order reduction techniques, and particularly in proper generalized decomposition (PGD) methods. We are aware that it looks often difficult to obtain a good PGD program and that there is a sort of steep learning curve. To overcome these difficulties, this book is thought to ease as much as possible the coding phase of PGD.

Many other books devoted to model order reduction in general, and PGD in particular, exist nowadays. But we strongly believe that this one covers aspects not fully considered in previous works on the topic.

Maybe the greatest advantage of PGD over other model reduction techniques, if any, relies in its ability of coping with parametric problems defined in high-dimensional *phase spaces*. This book begins with the most classical Poisson problem and soon moves to parametric problems in a wider sense. Among all the possible parametric problems, we have chosen some that can be considered *not classical*. Covered problems are not classical in the sense that we extend the concept of *parameter* far beyond the classical meaning of the word. Thus, we show that boundary conditions, and in particular loads, can be considered as parameters. In Chap. 3, we will show how the position of a load can be efficiently considered as a parameter under the PGD rationale, leading in fact to a very simple interactive program in which the user can play with a cantilever beam and see in real time its deformed configuration. Although the problem of obtaining a response surface for a moving load has traditionally been seen as inseparable or, in other words, nonreducible, we show that it can be effectively reduced under the PGD prism. In an offline phase of the methods functions or modes approximating the solution are computed, so as to allow in an online phase to obtain a response in real time.

In Chap. 4, we extend the previous development to the case of nonlinear problems, taking hyperelasticity as a model problem. Nonlinear problems continue to be a headache for model order reduction techniques, since they provoke the loss of the most part of the gains of model order reduction and every time the tangent stiffness matrix of the problem needs to be reassembled. In this chapter we show

how a very simple explicit linearization leads to a neat program, able to provide three-dimensional results under real-time constraints.

In Chap. 5 we turn the concept of parameter up a notch. Indeed, we show how initial and boundary-value problems can be effectively reduced under the PGD framework by considering the fields of initial conditions as parameters. But initial conditions are indeed magnitudes of infinite dimension, and therefore there is a need for subsequent reduction. After finite element discretization, a proper orthogonal decomposition is applied over some snapshots of problems similar to the one at hand. Then, with a minimal number of parameters, initial conditions can be considered effectively as new parameters of the model.

By taking solid dynamics as a model problem, we show that PGD gives a very practical response for initial and boundary-value problems. These approaches have rendered impressive gains in terms of computational cost, allowing for real-time applications infields such as virtual surgery, among others.

We are confident that the reader will find his or her problem of interest represented by any of the chosen examples and that the accompanying Matlab codes will make his or her life easier.

Zaragoza
December 2015

Elías Cueto
David González
Icíar Alfaro

Acknowledgments

We would like to thank Prof. Francisco (Paco) Chinesta for all these years of fruitful work and friendship. It has been a pleasure to work together. It is with him that we have met the joy of science. We also would like to thank our students, who worked hard towards the achievement of their doctoral degree. Our most sincere acknowledgment goes to Siamak Niroomandi, Carlos Quesada, and Diego Canales. Last but not least, it has been an honor and a pleasure to work with our colleagues and friends at Polytechnic University of Catalonia in Barcelona, Profs. Antonio Huerta and Pedro Díez, with whom we worked in two different projects related to PGD under the financial support of the Spanish Ministry of Economy and Competitiveness.

Contents

Chapter 1
Introduction

Young men should prove theorems, old men should write books.
—G.H. Hardy

Abstract This introductory chapter covers briefly the main idea of the book, how to code Proper Generalized Decomposition techniques with Matlab.

Proper Generalized Decomposition (PGD) has revolutionized many fields of applied sciences and engineering, and particularly the way we see parametric problems in a high dimensional setting. It has revealed how to obtain reduced-order models without the need for complex and costly computational experiments, typical of *a posteriori* model order reduction techniques such as the much better know Proper Orthogonal Decompositions (POD). These techniques, developed and re-discovered in many branches of science under different names such as Principal Component Analysis (PCA) [34], but also Karhunen-Loève transform [43, 48] in signal processing, also the Hotelling transform, Eckart-Young theorem, singular value decomposition (SVD), eigenvalue decomposition (EVD) in linear algebra, factor analysis, empirical orthogonal functions (EOF), empirical eigenfunction decomposition, empirical component analysis [49], quasiharmonic modes, and empirical modal analysis in structural dynamics. All these methods need for some snapshots, empirical realizations of the problem at hand under different values of the considered parameters. From these snapshots, the eigenmodes that contain most of the energy of their autocorrelation matrix are retained and employed as the best possible basis (given these realizations or snapshots) for subsequent simulations under different values of the parameters.

This approach (often referred to as *projection-based* methods, since the discrete equations are projected onto a reduced-order subspace in which the number of degrees of freedom of the problem is minimal) has, noteworthy, several disadvantages. In non-linear problems, for instance, the application of Newton algorithms for the linearization implies an update of the tangent stiffness matrix of the full problem, and

E. Cueto et al., *Proper Generalized Decompositions*, SpringerBriefs
in Applied Sciences and Technology, DOI 10.1007/978-3-319-29994-5_1

hence the loss of most of the time savings obtained through POD. Although several alternatives exist, such as the use of empirical interpolation methods, for instance, [14, 21, 53], or the use of perturbation techniques in conjunction with POD, [58], no definitive answer has been given to the model order reduction of non-linear problems.

The origin of Proper Generalized Decompositions, can be traced back to the so-called *radial loading* within the LATIN method [44] as a space-time separated representation in non-incremental structural mechanics solvers. Independently, Chinesta and coworkers in their seminal papers [6, 7] developed a method for the solution of non-Newtonian fluid models defined in high-dimensional phase spaces that were soon identified as a generalization of the work by P. Ladeveze. PGD is constructed indeed upon a very old idea, the method of separation of variables or Fourier method for partial differential equations. But the main novelty lies in the ability of PGD for the construction of sums of separated functions *a priori*, i.e., without any prior knowledge on the solution nor the need for costly computer experiments or snapshots. Indeed, a PGD approximation to the solution of a given PDE, say u, depending in principle of space, time and a number m of parameters, assumes a form

$$u(x,t,p_1,p_2,\ldots,p_m) \approx u^n = \sum_{i=1}^{n} F_x^i(x) \cdot F_t^i(t) \cdot F_1^i(p_1) \cdot F_2^i(p_2) \cdot \ldots \cdot F_m^i(p_m), \tag{1.1}$$

where the functions F_i^j are in principle unknown and p_i represent the parameters affecting the solution.

Briefly speaking, to determine these functions F_i^j a non-linear problem must be solved (independently of the character of the initial problem), since we look for one or more products of functions. Surprisingly or not, very simple techniques have demonstrated to provide very good results. Thus, for instance, in many references of the PGD bibliography (see [24, 29] for recents reviews on the field) greedy algorithms are employed such that one sum is computed at a time, and within each greedy step, naive linearization strategies such as fixed point iterations usually provide very good results.

The choice of an appropriate truncation level n deserves some comments. In fact, Eq. (1.1) would need, in the most general case, an infinite number of terms. To properly determine the number of terms n needed to obtain a certain level of accuracy, several possibilities exist. The most rigorous ones include the computation of error estimators, possibly based on engineering quantities of interest, see [3, 4, 19, 46, 52]. It is also worthy of mention that the PGD method, as stablished in Eq. (1.1) has proven convergence for elliptic problems [47].

This very simple approach has allowed to solve parametric problems in phase spaces of one hundred dimensions but, notably, it has provided new insights in the way we look at physical phenomena governed by partial differential equations. Thus, for instance, treating as parameters things that *a priori* are not parameters (such as loads, boundary conditions, initial conditions, ...) allows for a very efficient solution of different problems in engineering and applied sciences, that sometimes reaches real-time constraints by employing an off-line/on-line strategy.

Just to show some examples, PGD has allowed to construct efficient surgery simulators with haptic response, see Fig. 1.1, or to enable not-so-simple simulations on handheld, deployed devices such as smartphones or tablets, Fig. 1.2, to embed complex simulations on a simple web page (by employing simple java applets), see Fig. 1.3. The developed technique could even have important implications in augmented learning environments, opening the possibility to include real-time simulations on e-books, Fig. 1.4 [60].

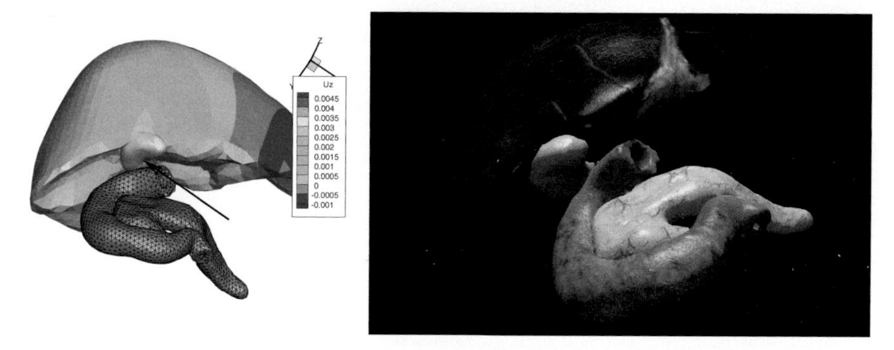

Fig. 1.1 An example of surgery simulator developed with the aid of PGD methods [57]

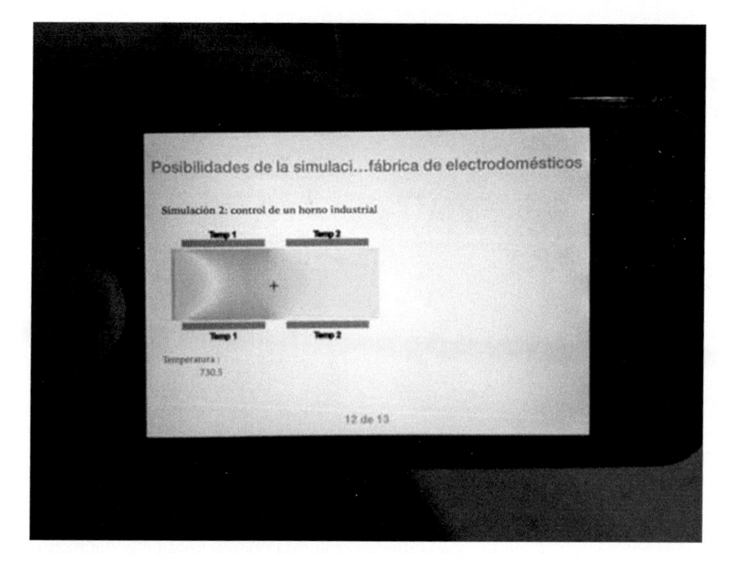

Fig. 1.2 Interactive simulation of an industrial furnace running on an iPhone [35]. The simulation could eventually be continuously fed by data streaming from sensors. It is what is known as dynamic, data-driven applications systems (DDDAS)

Fig. 1.3 Interactive palpation of a liver running a small java applet on a web page

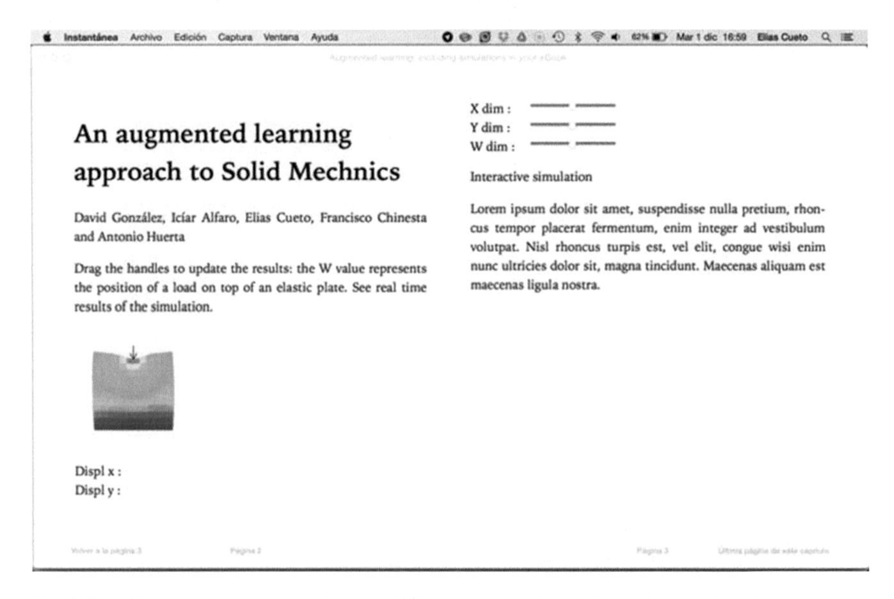

Fig. 1.4 PGD techniques open the possibility to embed real-time, interactive simulations on e-books, thus opening unprecedented possibilities for augmented learning environments [60]

In recent times, PGD has been applied to a wide variety of problems, showing its impressive ability to give appropriate responses in very different fields. Thus, for instance, one can mention the solution of Helmholtz equations [13, 51], geophysical problems [63], magnetostatics [42], real-time monitoring [2], Boltzmann and Fokker-Planck equations [22], gene regulatory networks [5, 30], contact problems [37, 40], the construction of response surfaces, virtual charts or computational vademecums [26, 61], multiscale problems [9, 23, 25, 32, 45], shape optimization [10], dynamic, data driven application systems (DDDAS) [36, 39], or virtual surgery [3, 33, 56, 59], to cite but a few of the more than 300 references available in the World of Science as of December, 2015. In addition, PGD is a quite intrusive method that can not, in general, be applied with the help of commercial finite element codes. Nevertheless, it has been efficiently coupled to existing techniques in a number of references, see, for instance, [8, 11].

However, from the programmer point of view, PGD has a clear barrier to entry. Even for experienced finite element programmers, PGD could appear as something complex, obscure, with many tricks to know in advance and difficult to understand. It is not the purpose of this *brief* to cover all the theory related to PGD, something already done in previous monographs, see [27, 28], for instance. Instead, this book is devoted to an easy and friendly introduction to PGD programming with one of the most popular languages for applied scientists and engineers, Matlab.

Chapter 2
To Begin With: PGD for Poisson Problems

It is one of the first duties of a professor, for example, in any subject, to exaggerate a litlle both the importance of his subject and his own importance in it

—G.H. Hardy, 1940

Abstract In this chapter we cover the detailed implementation of PGD methods for the simplest problem, the Poisson equation. Detailed code is provided and its results compared with data available in the bibliography.

2.1 Introduction

To begin with, let us consider one of the simplest problems, the Poisson equation. This problem was first analyzed in the original publication by Ammar et al. [6], still when the term PGD had not been coined.

Even if it is of little practical interest (very rarely we are interested in separating the space coordinates, unless very special cases such as plate and shell geometries [16] for instance), let us briefly recall the PGD method applied to a Poisson problem, still in two dimensions, for the ease of representation. In it, coordinates x and y have been separated with an eye towards the full comprehension of the method. In general, for non-separable (non parallelepipedic) domains, to separate the physical space is not possible nor even desirable. If you are, nevertheless forced to do it, maybe you could be interested in reading the reference [38].

E. Cueto et al., *Proper Generalized Decompositions*, SpringerBriefs
in Applied Sciences and Technology, DOI 10.1007/978-3-319-29994-5_2

2.2 The Poisson Problem

The D-dimensional Poisson equation writes

$$\Delta u = -f(x_1, x_2, \ldots, x_D), \tag{2.1}$$

where u is a scalar function of $(x_1, x_2, \ldots, x_D)$. Although for representation purposes we will restrict ourselves to the two-dimensional case, we consider here Eq. (2.1), defined in the domain $(x_1, x_2, \ldots, x_D) \in \Omega = (-L, +L)^D$ with vanishing essential boundary conditions. The general treatment of non-vanishing boundary conditions under the PGD framework needs for an special (although straightforward) treatment, deeply discussed from a practical point of view in [38]. We refer the interested reader to that reference.

Under the basic PGD assumption given by Eq. (1.1), we express the unknown solution field as a sum of separable functions, i.e.,

$$u(x_1, x_2, \ldots, x_D) = \sum_{j=1}^{\infty} \alpha_j \prod_{k=1}^{D} F_k^j(x_k),$$

where F_k^j is the jth basis function, with unit norm, which only depends on the kth coordinate.

This a priori infinite sum is then truncated (usually with the help of some error indicator, see [4, 19, 46, 52]) at a number (n) of approximation functions, i.e.:

$$u(x_1, x_2, \ldots, x_D) \approx \sum_{j=1}^{n} \alpha_j \prod_{k=1}^{D} F_k^j(x_k). \tag{2.2}$$

Note that originally, in [6], the separate functions F_k^j (which we later refer to as *modes* very often) were of unity norm. This is the origin of the α_j weight that accompanies each term j of the sum. While it is not strictly necessary to employ such unitary functions, the relative decay of the α_j weights with j gives a very intuitive notion on the convergence of the series.

Note also that, in Eq. (2.2) only one-dimensional functions F_k^j have been considered. The method is of course much more general than that, and a combination of functions defined in arbitrary dimensional spaces can be employed. Also the (in principle 1D) mesh employed for each function need not be uniform. h-refinements can be made along each dimension as needed.

The modes F_k^j at a given iteration of the method, j, and the α_j value need now to be computed. In the original paper [6] an algorithm was proposed that proceeded by

Step 1: *Projection of the solution in a discrete basis*

If we assume functions $F_k^j (\forall j \in [1, \dots, n]; \forall k \in [1, \dots, D])$ already known, coefficients α_j can be computed by introducing the approximation of u into the Galerkin variational formulation associated with Eq. (2.1):

$$\int_{\Omega} \nabla u^* \cdot \nabla u d\Omega = \int_{\Omega} u^* f \, d\Omega. \tag{2.3}$$

PGD methods assume a separated representation of both u and u^*:

$$u(x_1, x_2, \dots, x_D) = \sum_{j=1}^{n} \alpha_j \prod_{k=1}^{D} F_k^j(x_k), \tag{2.4}$$

and

$$u^*(x_1, x_2, \dots, x_D) = \sum_{j=1}^{n} \alpha_j^* \prod_{k=1}^{D} F_k^j(x_k).$$

By introducing both in the weak form of the problem, Eq. (2.3), we arrive at

$$\int_{\Omega} \nabla \left(\sum_{j=1}^{n} \alpha_j^* \prod_{k=1}^{D} F_k^j(x_k) \right) \cdot \nabla \left(\sum_{j=1}^{n} \alpha_j \prod_{k=1}^{D} F_k^j(x_k) \right) d\Omega$$
$$= \int_{\Omega} \left(\sum_{j=1}^{n} \alpha_j^* \prod_{k=1}^{D} F_k^j(x_k) \right) f \, d\Omega \tag{2.5}$$

We assume also that the source term $f(x_1, \cdots, x_D)$ admits a separated representation

$$f(x_1, \cdots, x_D) \approx \sum_{h=1}^{m} \prod_{k=1}^{D} f_k^h(x_k),$$

for a sufficiently low number of terms m. If it is not the case, a simple singular value decomposition would in general suffice, maybe in high dimensions (HOSVD) [38]. PGD could equally be employed to this end, by applying it on the identity operator, see [28].

Equation (2.5) involves integrals of products involving D different functions, each one defined in a different coordinate. Let $\prod_{k=1}^{D} g_k(x_k)$ be one of these functions to be integrated. One of the key ingredients of PGD is that the integral over Ω can be performed by integrating each function along its definition interval and then multiplying the D computed integrals according to:

$$\int_{\Omega} \prod_{k=1}^{D} g_k(x_k)\, d\Omega = \prod_{k=1}^{D} \int_{-L}^{L} g_k(x_k) dx_k.$$

This constitutes an essential feature of PGD that makes it possible to solve problems defined in high dimensional spaces.

Since u^* represents an admissible variation of the solution u, the weights α_j^* are arbitrary, too (very much like nodal coefficients of admissible variations in FEM). Thus, Eq. (2.5) allows to compute the n approximation coefficients α_j, solving the resulting linear system of size $n \times n$. This problem is linear and moreover rarely exceeds the order of some tens of degrees of freedom. Thus, even if the resulting coefficient matrix is densely populated, the time required for its solution is negligible with respect to the one required for performing the approximation basis enrichment (step 3).

Step 2: *Checking convergence*

From the solution of u at iteration n given by Eq. (2.4) the residual Re related to Eq. (2.1) can be computed:

$$Re = \frac{\sqrt{\int_{\Omega} \left(\Delta u + f(x_1, \ldots, x_D)\right)^2}}{u}. \tag{2.6}$$

By fixing a tolerance $Re < \epsilon$, the iteration process can be stopped, thus providing the solution $u(x_1, \ldots, x_D)$.

As per the weak form of the problem, the integral in Eq. (2.6) can be written as the product of one-dimensional integrals by performing a separated representation of the square of the residual.

Step 3: *Enrichment of the approximation basis*

If the stopping criterion has not yet be accomplished, the PGD approximation can be enriched by adding a new functional product $\prod_{k=1}^{D} F_k^{(n+1)}(x_k)$. To this end, the non-linear Galerkin variational formulation related to Eq. (2.1) is then solved:

$$\int_{\Omega} \nabla u^* \cdot \nabla u d\Omega = \int_{\Omega} u^* f \, d\Omega,$$

using the approximation of u given by:

$$u(x_1, x_2, \ldots, x_D) = \sum_{j=1}^{n} \alpha_j \prod_{k=1}^{D} F_k^j(x_k) + \prod_{k=1}^{D} R_k(x_k).$$

Identically, the test function has the form

$$u^*(x_1, x_2, \ldots, x_D) = R_1^*(x_1)\cdot R_2(x_2)\cdot\ldots\cdot R_D(x_D)+\ldots+R_1(x_1)\cdot R_2(x_2)\cdot\ldots\cdot R_D^*(x_D),$$

by simply applying the rules of variational calculus.

This leads finally to a non-linear variational problem (note that we seek a product of functions expressed in a one dimensional finite element basis), whose solution allows to compute the D sought functions $R_k(x_k)$. Functions $F_k^{(n+1)}(x_k)$ need finally to be normalized.

To solve this problem we introduce a discretization of those functions $R_k(x_k)$. Each one of these functions is approximated using a 1D finite element description. If we assume than p_k nodes are used to construct the interpolation of function $R_k(x_k)$ in the interval $[-L, L]$, then the size of the resulting discrete non-linear problem is $\sum_{k=1}^{k=D} p_k$. The price to pay for avoiding a whole mesh in the multidimensional domain is the solution of a non-linear problem. However, even in high dimensions the size of the non-linear problems remains moderate and no particular difficulties have been found in its solution up to hundreds dimensions. Concerning the computation time, even when the non-linear solver converges quickly, this step consumes the main part of the global computing time.

Different non-linear solvers have been analyzed: Newton or one based on an alternating directions scheme. In this work the last strategy was retained. Thus, in the enrichment step, function $R_1^{s+1}(x_1)$ is updated by assuming known all the others functions (given at the previous iteration of the non-linear solver $R_2^s(x_2), \cdots, R_D^s(x_D)$). Then, functions $R_1^{s+1}(x_1), R_3^s(x_3), \cdots, R_D^s(x_D)$ are assumed known for updating function $R_2^{s+1}(x_2)$, and so on until updating the last function $R_D^{s+1}(x_D)$. Now the convergence is checked by calculating $\sum_{i=1}^{i=D} R_i^{s+1}(x_i) - R_i^s(x_i)^2$. If this norm is small enough we can define the functions $F_k^{(n+1)}(x_k)$ by normalizing the functions $R_1, R_2, \ldots, R_D$ and come back to step 1. On the contrary, if this norm is not small enough, a new iteration of the non-linear solver should be performed by updating functions $R_i^{s+2}(x_i)$, $i = 1, \cdots, D$ and then checking again the convergence. Despite its simplicity, our experience proves that this strategy is in fact very robust.

We must recall that the technique that we proposed in the papers just referred, is not a universal strategy able to solve any kind of multidimensional partial differential equation (PDE). Thus, the efficient application of the technique that we just described requires the separability of all the fields involved in the model. Obviously, this separability is not always possible because some functions need a tremendous number of sums. On the other hand, even when the field is separable (one could perform this separation by invoking for example the SVD or the multidimensional SVD) the finite sums decomposition of general multidimensional functions is not realistic because the amount of memory needed for storing the discrete form of such functions before applying the multidimensional SVD.

In many physical models (see for example [6, 25]) a fully separation (consisting of a sum of products of one-dimensional functions) could not be envisaged from a practical point of view. Thus, a better approximation lies in writing

$$u(\boldsymbol{x}_1, \cdots, \boldsymbol{x}_d) \approx \sum_{i=1}^{i=N} F_1^i(\boldsymbol{x}_1) \times \cdots \times F_D^i(\boldsymbol{x}_d)$$

where the different functions taking part in the finite sums decomposition are defined in spaces of moderate dimensions, that is $\boldsymbol{x}_i \in \Omega_i \subset \mathbb{R}^{q_i}$, where usually $q_i = 1, 2$ or 3.

2.3 Matrix Structure of the Problem

The approximation to u given by Eq. (2.4) can indeed be further simplified by assuming

$$u(x_1, x_2, \ldots, x_D) = \sum_{j=1}^{n} \prod_{k=1}^{D} F_k^j(x_k), \tag{2.7}$$

i.e., there is no need to assume unit-normed functions and a weighting parameter α_j in the approximations of u. This was the initial approach followed in [6, 7], but we soon realized that it is equally possible to compute directly functions F_k without the need to enforce its unity norm, nor the projection stage of the algorithm presented before.

Consider, for simplicity, a two-dimensional code, although its extension to an arbitrary number of dimensions is strightforward. In it, functions F^i (we are going to rename them now by their two-dimensional counterparts $F^i(x)$ and $G^i(y)$) are approximated by employing (linear in this case) finite elements, so that, at iteration n, the i-th sum of the approximation will be given by

$$u^i(x, y) = \left[\boldsymbol{N}^T \boldsymbol{F}_i \boldsymbol{M}^T \boldsymbol{G}_i\right], \tag{2.8}$$

with $\boldsymbol{N}$ and $\boldsymbol{M}$ the vectors containing the values of finite element shape functions at integration points and $\boldsymbol{F}_i$ and $\boldsymbol{G}_i$ the vectors of nodal values at the FE mesh for the functions $F^i(x)$ and $G^i(y)$, respectively. In the code included below we assume identical approximation along x and y directions so that only a matrix $\boldsymbol{N} = \boldsymbol{M}$ will be necessary.

The same is necessary for the computation of the gradient terms arising in Eq. (2.3),

$$\begin{bmatrix} \frac{\partial u}{\partial x} & \frac{\partial u}{\partial y} \end{bmatrix}^T = \begin{bmatrix} d\boldsymbol{N}^T \boldsymbol{F}_1 \boldsymbol{M}^T \boldsymbol{G}_1 \; d\boldsymbol{N}^T \boldsymbol{F}_2 \boldsymbol{M}^T \boldsymbol{G}_2 \ldots d\boldsymbol{N}^T \boldsymbol{F}_n \boldsymbol{M}^T \boldsymbol{G}_n \\ \boldsymbol{N}^T \boldsymbol{F}_1 d\boldsymbol{M}^T \boldsymbol{G}_1 \; \boldsymbol{N}^T \boldsymbol{F}_2 d\boldsymbol{M}^T \boldsymbol{G}_2 \ldots \boldsymbol{N}^T \boldsymbol{F}_n d\boldsymbol{M}^T \boldsymbol{G}_n \end{bmatrix},$$

where $d\boldsymbol{N}$ and $d\boldsymbol{M}$ represent the vector containing the value of shape function's derivatives at Gauss points. A similar expression can be envisaged both for u^* and ∇u^*, while in this case the nodal values of functions $\boldsymbol{F}_i^*$ and $\boldsymbol{G}_i^*$ are arbitrary, as it is well known from standard finite element theories.

When we look for a new term in the approximation, we assume that

$$u(x, y) = \sum_{i=1}^{n} F^i(x)G^i(y) + R(x)S(y), \tag{2.9}$$

while

$$u^*(x, y) = R^*(x)S(y) + R(x)S^*(y).$$

The new, enhanced, expression of the gradients will be

$$\begin{bmatrix} \frac{\partial u}{\partial x} & \frac{\partial u}{\partial y} \end{bmatrix}^T = \sum_i \begin{bmatrix} d\boldsymbol{N}^T \boldsymbol{F}_i \boldsymbol{M}^T \boldsymbol{G}_i \\ \boldsymbol{N}^T \boldsymbol{F}_i d\boldsymbol{M}^T \boldsymbol{G}_i \end{bmatrix} + \begin{bmatrix} \boldsymbol{M}^T \boldsymbol{S} d\boldsymbol{N}^T & \boldsymbol{0} \\ \boldsymbol{0} & \boldsymbol{N}^T \boldsymbol{R} d\boldsymbol{M}^T \end{bmatrix} \begin{bmatrix} \boldsymbol{R} \\ \boldsymbol{S} \end{bmatrix}$$
$$= \sum_i \boldsymbol{D}_i + \boldsymbol{E} \begin{bmatrix} \boldsymbol{R} \\ \boldsymbol{S} \end{bmatrix},$$

and, similarly,

$$u^*(x, y) = \begin{bmatrix} \boldsymbol{R}^T & \boldsymbol{S}^T \end{bmatrix} \begin{bmatrix} \boldsymbol{S}^T \boldsymbol{M} \boldsymbol{N} \\ \boldsymbol{R}^T \boldsymbol{N} \boldsymbol{M} \end{bmatrix}.$$

The same must be done for

$$\begin{bmatrix} \frac{\partial u^*}{\partial x} & \frac{\partial u^*}{\partial y} \end{bmatrix} = \begin{bmatrix} \boldsymbol{R}^{*T} & \boldsymbol{S}^{*T} \end{bmatrix} \begin{bmatrix} \boldsymbol{S}^T \boldsymbol{M} d\boldsymbol{N} & \boldsymbol{S}^T d\boldsymbol{M} \boldsymbol{N} \\ \boldsymbol{R}^T d\boldsymbol{N} \boldsymbol{M} & \boldsymbol{R}^T \boldsymbol{N} d\boldsymbol{M} \end{bmatrix} = \begin{bmatrix} \boldsymbol{R}^{*T} & \boldsymbol{S}^{*T} \end{bmatrix} \boldsymbol{F}^T,$$

and for the source term, by assuming that

$$f(x, y) \approx \sum_h a^h(x) b^h(y).$$

Once all this matrices have been substituted into the weak form of the problem, Eq. (2.3) ,we arrive at

$$\int_\Omega \begin{bmatrix} \boldsymbol{R}^{*T} & \boldsymbol{S}^{*T} \end{bmatrix} \sum_i \boldsymbol{F}^T \boldsymbol{D}_i d\Omega + \int_\Omega \begin{bmatrix} \boldsymbol{R}^{*T} & \boldsymbol{S}^{*T} \end{bmatrix} \boldsymbol{F}^T \boldsymbol{E} \begin{bmatrix} \boldsymbol{R}^T \\ \boldsymbol{S}^T \end{bmatrix} d\Omega$$
$$= \sum_h \int_\Omega \begin{bmatrix} \boldsymbol{R}^{*T} & \boldsymbol{S}^{*T} \end{bmatrix} \begin{bmatrix} \boldsymbol{S}^T \boldsymbol{M} b^h(y) \boldsymbol{N} a^h(x) \\ \boldsymbol{R}^T \boldsymbol{N} a^h(x) \boldsymbol{M} b^h(y) \end{bmatrix} d\Omega. \tag{2.10}$$

These integrals in Ω can in fact be separated (since every term is) into a sequence of integrals along x and y coordinates. The resulting terms for matrices $\boldsymbol{F}^T\boldsymbol{D}_i$ and $\boldsymbol{F}^T\boldsymbol{E}$ will involve a repeated evaluation of four terms, namely,

$$\int_{x=-L}^{x=+L} d\boldsymbol{N} d\boldsymbol{N}^T dx \text{ and } \int_{y=-L}^{y=+L} d\boldsymbol{M} d\boldsymbol{M}^T dy \tag{2.11}$$

and

$$\int_{x=-L}^{x=+L} \boldsymbol{N}\boldsymbol{N}^T dx \text{ and } \int_{y=-L}^{y=+L} \boldsymbol{M}\boldsymbol{M}^T dy, \tag{2.12}$$

which are referred to as `p1` and `p2`, respectively, in routine `elemstiff` (see the call `[Km{i1},Mm{i1}] = elemstiff(coor{i1})` in the main file of the code). The code below in fact assumes that $\boldsymbol{N} = \boldsymbol{M}$, since equal partitions are made along x and y directions. For instance, the term 11 of the integration of matrix $\boldsymbol{F}^T\boldsymbol{E}$ has the form,

$$\int_{\Omega} (\boldsymbol{F}^T\boldsymbol{E})_{11} d\Omega = \left(\int_{x=-L}^{x=+L} d\boldsymbol{N} d\boldsymbol{N}^T dx\right) \boldsymbol{S}^T \left(\int_{y=-L}^{y=+L} \boldsymbol{M}\boldsymbol{M}^T dy\right) \boldsymbol{S}. \tag{2.13}$$

Similarly, the right-hand-side term in Eq. (2.10) has the form,

$$\sum_{h=1}^{m} \begin{bmatrix} \boldsymbol{S}^T \left(\int_{y=-L}^{y=+L} \boldsymbol{M} b^h(y) dy\right) \left(\int_{x=-L}^{x=+L} \boldsymbol{N} a^h(x) dx\right) \\ \boldsymbol{R}^T \left(\int_{x=-L}^{x=+L} \boldsymbol{N} a^h(x) dx\right) \left(\int_{y=-L}^{y=+L} \boldsymbol{M} b^h(y) dy\right) \end{bmatrix}. \tag{2.14}$$

The problem, finally, renders Eq. (2.3) in a matrix form that can be simplified, after invoking the arbitrariness of $\boldsymbol{R}^{*T}$ and $\boldsymbol{S}^{*T}$, to

$$\boldsymbol{V}_1(\boldsymbol{R}, \boldsymbol{S}) + \boldsymbol{K}(\boldsymbol{R}, \boldsymbol{S}) \begin{bmatrix} \boldsymbol{R} \\ \boldsymbol{S} \end{bmatrix} = \boldsymbol{V}_2(\boldsymbol{R}, \boldsymbol{S}),$$

which is more easily recognized if we write it in the form

$$\boldsymbol{K}(\boldsymbol{R}, \boldsymbol{S}) \begin{bmatrix} \boldsymbol{R} \\ \boldsymbol{S} \end{bmatrix} = \boldsymbol{V}_2(\boldsymbol{R}, \boldsymbol{S}) - \boldsymbol{V}_1(\boldsymbol{R}, \boldsymbol{S}) = \boldsymbol{V}(\boldsymbol{R}, \boldsymbol{S}). \tag{2.15}$$

It is important to note that the problem in Eq. (2.15) is non-linear, since we look for functions R and S, but both appear multiplied to each other in Eq. (2.9). You can choose your favorite linearization procedure (Newton methods, for instance). In our experience, fixed-point, alternated directions algorithms render excellent results and, despite their general lack of convergence, this is rarely found in practice.

2.4 Matlab Code for the Poisson Problem

The code main file is called `main.m`, of course! Its content is reproduced below.

```
%
%                    PGD Code for Poisson problems
%                  D. Gonzalez, I. Alfaro, E. Cueto
%                      Universidad de Zaragoza
%                        AMB-I3A Dec 2015
%
clear all; close all; clc;
%
% VARIABLES
%
ndim = 2; nn = 40.*ones(ndim,1); % # of Dimensions, # of Elements
num_max_iter = 15; % Max. # of summands for the approach
TOL = 1.0E-4; npg = 2; % Tolerance, Gauss Points
coor = cell(ndim,1); % Coordinates in each direction
L0 = -1.*ones(ndim,1);
L1 = ones(ndim,1); % Geometry (min,max coordinates)
for i1=1:ndim,
    coor{i1} = linspace(L0(i1),L1(i1),nn(i1));
end
%
% ALLOCATION OF IMPORTANT MATRICES
%
Km = cell(ndim,1); % "Stiffness" matrix \int dN dN dx, Eq. (2.10)
Fv = cell(ndim,1); % R and S sought enrichment functions
Mm = cell(ndim,1); % "Mass" matrix \int N N dx, Eq. (2.11)
V = cell(ndim,1);  % Source term in separated form
%
% COMPUTING STIFFNESS AND MASS MATRICES ALONG EACH DIRECTION
%
for i1=1:ndim,
    [Km{i1},Mm{i1}] = elemstiff(coor{i1});
end
%
% SOURCE TERM IN SEPARATED FORM
%
% Let us begin by a separable expression. Evaluation of Eq. (2.13)
Ch{1,1} = @(x) cos(2*pi*x); Ch{2,1} = @(y) sin(2*pi*y);
% Try this new source term by yourself by simply uncommenting next 2 lines!
% Ch{1,1} = @(x) x.*x; Ch{1,2} = @(x) -1.0+0.0*x;
% Ch{2,1} = @(y) 1.0+0.0*y; Ch{2,2} = @(y) y.*y;
for j1=1:ndim
    for k1=1:size(Ch,2)
        V{j1}(:,k1) = Ch{j1,k1}(coor{j1});
    end
    % Although in this case we have a closed-form expression for the source
    % term, in general we know its nodal values.
    V{j1} = Mm{j1}*V{j1};
end
%
% BOUNDARY CONDITIONS
%
CC = cell(ndim,1);
for i1=1:ndim,
    IndBcnode{i1} = [1 numel(coor{i1})];
end
for i1=1:ndim,
    CC{i1} = setxor(IndBcnode{i1},[1:numel(coor{i1})])';
end
%
% ENRICHMENT OF THE APPROXIMATION, LOOKING FOR R AND S
%
```

```
num_iter = 0; iter = zeros(1); Aprt = 0; Error_iter = 1.0;
while Error_iter>TOL && num_iter<num_max_iter
    num_iter = num_iter + 1; R0 = cell(ndim,1);
    for i1=1:ndim
      % Initial guess for R and S.
      % It works equally well by choosing something random.
      R0{i1} = ones(numel(coor{i1}),1);
     % We impose that initial guess for functions R and S verify
     % homogeneous essential boundary conditions.
      R0{i1}(IndBcnode{i1}) = 0;
    end
    %
    % ENRICHMENT STEP
    %
    [R,iter(num_iter)] = enrichment(Km,Mm,V,num_iter,Fv,R0,CC,TOL);
    for i1=1:ndim, Fv{i1}(:,num_iter) = R{i1}; end % R (S) is valid, add it
    %
    % STOPPING CRITERION
    %
    Error_iter = 1.0;
    % One possible criterion is to stop when the norm of the new sum is
    % negligible wit the pair of functions with the maximum norm
    for i1=1:ndim, Error_iter = Error_iter.*norm(Fv{i1}(:,num_iter)); end
    Aprt = max(Aprt,sqrt(Error_iter));
    Error_iter = sqrt(Error_iter)/Aprt;
    fprintf(1,'%dst summand in %d iterations with a weight of %f\n',...
        num_iter,iter(num_iter),sqrt(Error_iter));
end
num_iter = num_iter - 1;% the last sum was negligible, we discard it.
fprintf(1,'PGD Process exited normally\n\n');
save('WorkSpacePGD_Basic.mat');
%
% POST-PROCESSING
%
for i1=1:ndim
    figure;
    plot(coor{i1},Fv{i1}(:,1:num_iter));
end
figure;
if ndim==2
    surf(coor{1},coor{2},Fv{2}*Fv{1}');
end
```

The source code reproduce before includes a call to a function called `elemstiff` that, obviously, provides with the stiffness matrix for each 1D element in the problem. It is reproduced below:

```
function [p1,p2] = elemstiff(coor)
% function [p1,p2] = elemstiff(coor)
% For the coordinates coor, obtains p1 (Stiffness) and p2(Mass) matrices
% Universidad de Zaragoza - 2015

nen = numel(coor); p1 = zeros(nen); p2 = zeros(nen); %  p3 = zeros(nen,1);
X = coor(1:nen-1)'; % Left coordinate of the elements
Y = coor(2:nen)'; % Right coordinate of the elements
L = Y - X; % Length of the elements
sg = [-0.57735027, 0.57735027]; wg = [1, 1]; % Gauss and weight points
npg = numel(sg);
for i1=1:nen-1
    c = zeros(1,npg);
    N = zeros(nen,npg);
    dN = zeros(nen,npg);
    c(1,:) = 0.5.*(1.0-sg).*X(i1) + 0.5.*(1.0+sg).*Y(i1);
    N(i1+1,:) = (c(1,:)-X(i1))./L(i1);
    N(i1,:) = (Y(i1)-c(1,:))./L(i1);
```

```
        dN(i1+1,:) = ones(1,npg)./L(i1);
        dN(i1,:) = -dN(i1+1,:);
        for j1=1:npg
            p1 = p1 + dN(:,j1)*dN(:,j1)'*0.5.*wg(j1).*L(i1); % dNůdN
            p2 = p2 + N(:,j1)*N(:,j1)'.*0.5.*wg(j1).*L(i1); % NůN
        end
end
return
```

Once the sequence of 1D problems has lead us to a new term in the PGD approximation, we check if it is enough for the prescribed accuracy. If not, we add a new couple of functions in the `enrichment` function:

```
function [R,iter] = enrichment(K,M,V,num_iter,FV,R,CC,TOL)
% function [R,iter] = enrichment(K,M,V,num_iter,FV,R,CC,TOL)
% Compute a new sumand by fixed-point algorithm using PGD
% Universidad de Zaragoza, 2015
iter = 1;
mxit = 25; % # of possible iterations for the fixed point algorithm
error = 1.0e8; % Initialization
ndim = size(FV,1); % Number of Variables
%
% FIXED POINT ALGORITHM
%
while abs(error)>TOL
    Raux = R; % Remember: R is a cell containing both R and S
    for i1=1:ndim % Alternating between R and S
        matrix = zeros(numel(R{i1})); % K matrix in Eq. (2.14)
        source = zeros(size(R{i1},1),1);% V2-V1 in Eq. (2.14)
        %
        % COMPUTING K MATRIX Eq (2.14)
        %
        for i2=1:ndim % Products in sum = ndim (2 in this case)
            FTE = 1.0; % Computing F^T E in Eq. (2.8)
            % Remember: K = \int dN dN dx and M = \int N N dx
            for i3=1:ndim
                if i3==i2
                    if i3==i1
                        FTE = FTE.*K{i3};
                    else
                        FTE = FTE.*(R{i3}'*K{i3}*R{i3});
                    end
                else
                    if i3==i1
                        FTE = FTE.*M{i3};
                    else
                        FTE = FTE.*(R{i3}'*M{i3}*R{i3});
                    end
                end
            end
            matrix = matrix + FTE;
        end
        %
        % COMPUTING V2 in Eq. (2.14)
        %
        for j1=1:size(V{i1},2) % Number functions of the source
            V2 = 1.0;
            for i2=1:ndim
                if i2==i1
                    V2 = V2.*V{i2}(:,j1);
                else
                    V2 = V2.*(R{i2}'*V{i2}(:,j1));
                end
            end
            source = source + V2;
```

```
        end
        %
        % COMPUTING V1 in Eq. (2.14)
        %
        for j1=1:num_iter-1
            for i2=1:ndim % Terms in sum
                FTD = 1.0; % COMPUTING F^T D in Eq. (2.9)
                for i3=1:ndim
                    if i3==i2
                        if i3==i1
                            FTD = FTD.*(K{i3}*FV{i3}(:,j1));
                        else
                            FTD = FTD.*(R{i3}'*K{i3}*FV{i3}(:,j1));
                        end
                    else
                        if i3==i1
                            FTD = FTD.*(M{i3}*FV{i3}(:,j1));
                        else
                            FTD = FTD.*(R{i3}'*M{i3}*FV{i3}(:,j1));
                        end
                    end
                end
                source = source - FTD; % Note that source = V2-V1
            end
        end
        %
        % SOLVE Eq. (2.14) FOR EACH DIRECTION
        %
        R{i1}(CC{i1}) = matrix(CC{i1},CC{i1})\source(CC{i1});
        % We normalize S. R takes care of the alpha constant in Eq. (2.2)
        if i1~=1, R{i1} = R{i1}./norm(R{i1}); end
    end
    % If two successive Rs are too similar, we stop
    error = 0;
    for j1=1:ndim
        error = error + norm(Raux{j1}-R{j1});
    end
    error = sqrt(error);
    iter = iter + 1;
    if iter == mxit % If we reach the max # of iterations, we exit
        return;
    end
end
return
```

After executing this code in your own Matlab client, it provides you with the following figures. In Fig. 2.1 the obtained solution for the Poisson problem is show. Since the source term $\cos(2\pi x)\sin(2\pi y)$ is separable, the code provides the solution with one only term in the PGD sum. Of course, this is not always the case (indeed, it is almost never the case!). The two functions obtained whose multiplication gives the bi-dimensional solution are plotted in Fig. 2.2. Actually, both resemble very much to (the finite element approximation of) the cos and sin functions, respectively.

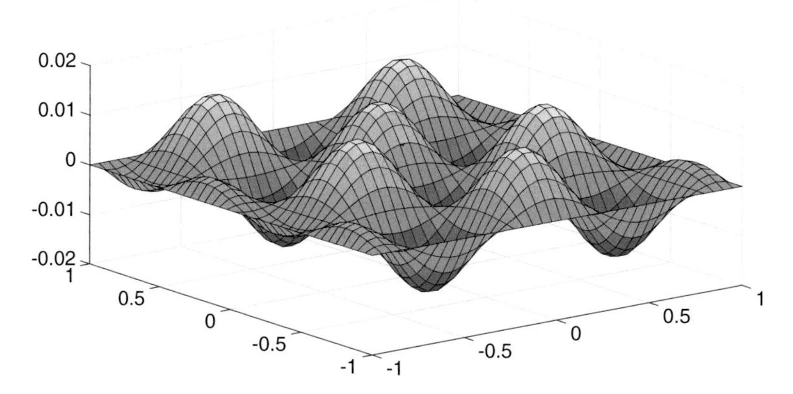

Fig. 2.1 Solution to the poisson problem, as given by the PGD code

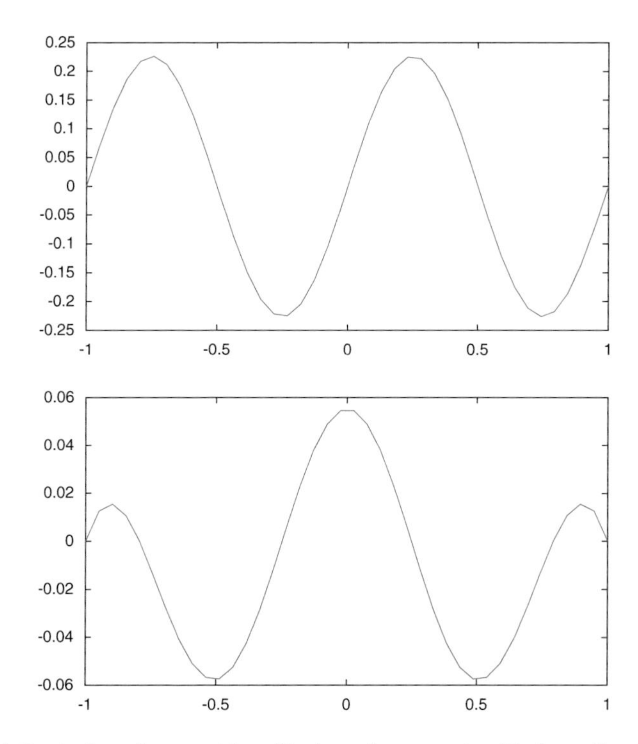

Fig. 2.2 Solution to the poisson problem. Single modes encountered in the x-direction (*top*) and y-direction (*bottom*)

Chapter 3
Parametric Problems

We have a habit in writing articles published in scientific journals to make the work as finished as possible, to cover all the tracks, to not worry about the blind alleys or to describe how you had the wrong idea first, and so on. So there isn't any place to publish, in a dignified manner, what you actually did in order to get to do the work.

—Richard Feynmann

Abstract This chapter develops the application of PGD methods for parametric problems, their natural field of application.

Parametric problems constitute perhaps the most relevant application of Proper Generalized Decomposition methods. Initially aimed at solving problems defined within a high dimensional phase space [6], PGD soon revealed an impressive ability to solve in the same setting parametric problems by just considering parameters as new dimensions, a sort of parametric phase space [27].

In this chapter we explore precisely these capabilities. Since they have also been included in a number of previous references, notably in former books (see [28], Chap. 5), we focus here on a particular instance of parametric problems: that of moving loads.

3.1 A Particularly Challenging Problem: A Moving Load as a Parameter

An *influence line* is a graphical representation of a given magnitude (a bending moment, for instance) at a given point of a beam caused by a unit load moving along the span of that beam. This concept has helped engineers through the years to design beam structures in an efficient manner. One can imagine, however, an extension of the concept of influence line to general, three-dimensional solids. This

E. Cueto et al., *Proper Generalized Decompositions*, SpringerBriefs
in Applied Sciences and Technology, DOI 10.1007/978-3-319-29994-5_3

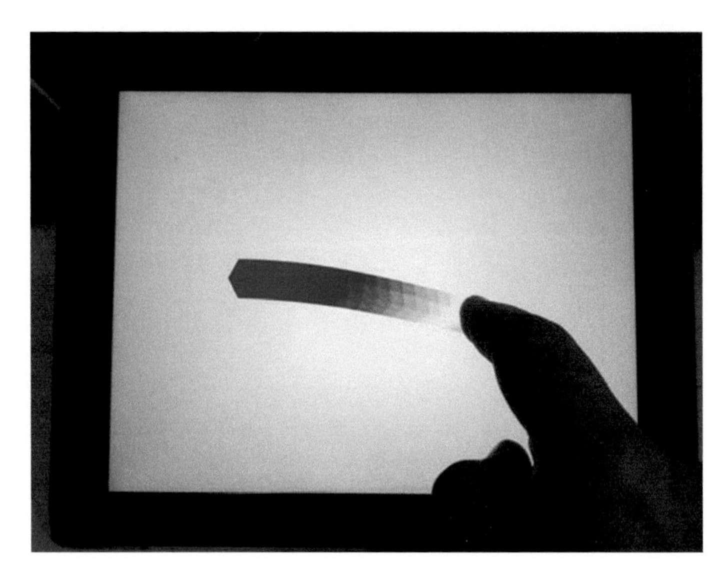

Fig. 3.1 An application of the influence line concept for a hyperelastic beam. You can play interactively with the beam by just placing the load with your finger on a tablet. A web-based version of this same app can be reached at http://amb.unizar.es/barra03.htm

would give rise to a sort of *response surface* in which particularizing the parameter (the position of the load) provides in an immediate way the response of the solid (its deformed configuration, in this case). This sort of influence line has potentially many applications in science and engineering for real-time simulation in fields such as computational surgery [55], decision taking, or even augmented learning [60], see Fig. 3.1.

In fact, this problem has been frequently thought of as *non separable*, i.e., that the number of modes needed to express the solution is so big, that no gain is obtained by applying any kind of model order reduction technique and therefore it is better to simply simulate it in a straightforward manner, by finite element methods or any other numerical technique of your choice.

However, it can be easily found that this is not true. Consider the influence line sketched in Fig. 3.2. We consider a clamped beam with a moving load and try to compute its deformed configuration for the load acting at any point. In fact, by applying Proper Orthogonal Decomposition techniques to the results, it can readily be seen that the number of modes or shape functions needed to express this solution $v = v(x, s)$ is in fact limited, see Fig. 3.3. Here, v denotes here the vertical displacement of the beam, x the particular point of the beam in which we want to know this displacement and s the position of the load,

As can be noticed from Fig. 3.3 (left), the eigenvalues decrease fast after a reasonable number of eigenmodes. Thus, even if we need to consider every possible load position, the number of functions needed to express the parametric solution is in fact reasonably limited.

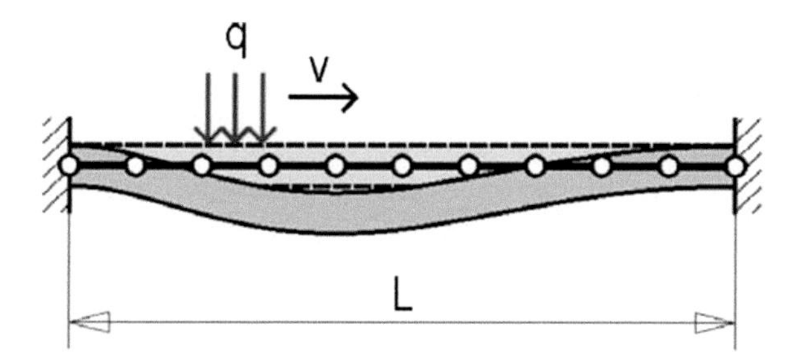

Fig. 3.2 A clamped beam with a moving load

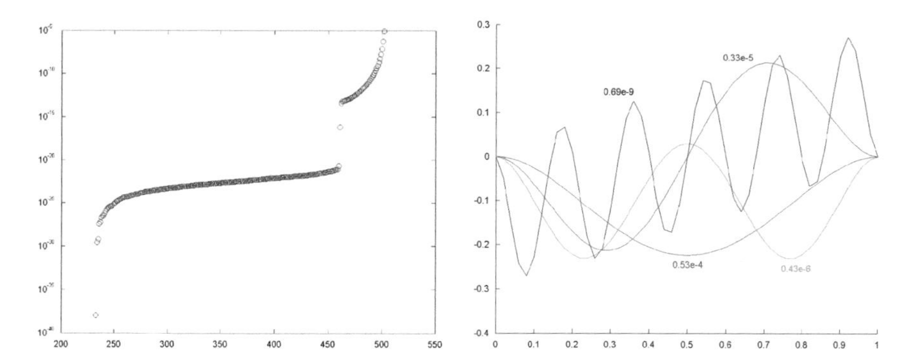

Fig. 3.3 (*Left*) resulting eigenvalues for the clamped beam problem and (*right*) first modes of the solution

3.2 The Problem Under the PGD Formalism

As will be noticed, under the PGD formalism, the *influence line* problem becomes simple. We generalize this problem so as to find the displacement field $\boldsymbol{u}(x, y, z)$ at any point of a three-dimensional solid Ω, for any load position $\boldsymbol{s} \in \bar{\Gamma} \subset \partial\Omega$.

Under these assumptions, the weak form of the problem will result after multiplying the strong form by an arbitrary test function $\boldsymbol{u}^*$ and integrating over the region occupied by the solid, Ω, and the portion of its boundary which is accessible to the load, $\bar{\Gamma}$. It therefore consists in finding the displacement $\boldsymbol{u} \in \mathcal{H}^1$ such that for all $\boldsymbol{u}^* \in \mathcal{H}^1_0$:

$$\int_{\bar{\Gamma}} \int_{\Omega} \nabla_s \boldsymbol{u}^* : \boldsymbol{\sigma} d\Omega d\bar{\Gamma} = \int_{\bar{\Gamma}} \int_{\Gamma_{t2}} \boldsymbol{u}^* \cdot \boldsymbol{t} d\Gamma d\bar{\Gamma} \tag{3.1}$$

where $\Gamma = \Gamma_u \cup \Gamma_t$ represents the essential and natural portions of the boundary, and where $\Gamma_t = \Gamma_{t1} \cup \Gamma_{t2}$, i.e., regions of homogeneous and non-homogeneous, respectively, natural boundary conditions. We assume, for simplicity of the exposition, that the load is of unity module and acts along the vertical axis: $\boldsymbol{t} = \boldsymbol{e}_k \cdot \delta(\boldsymbol{x} - \boldsymbol{s})$, where δ represents the Dirac-delta function and $\boldsymbol{e}_k$ the unit vector along the z-coordinate axis.

The Dirac-delta term needs to be approximated by a truncated series of separable functions in the spirit of the PGD method, i.e.,

$$t_j \approx \sum_{i=1}^{m} f_j^i(\boldsymbol{x}) g_j^i(\boldsymbol{s}) \tag{3.2}$$

where m represents the order of truncation and f_j^i, g_j^i represent the j-th component of vectorial functions in space and boundary position, respectively.

The PGD approach to the problem consists in finding, in a greedy way, a finite sum of separable functions to approach the solution. Assuming that, at iteration n of this procedure, we have converged to such an approximation, we have

$$u_j^n(\boldsymbol{x}, \boldsymbol{s}) = \sum_{k=1}^{n} F_j^k(\boldsymbol{x}) \cdot G_j^k(\boldsymbol{s}), \tag{3.3}$$

where the term u_j refers to the jth component of the displacement vector, $j = 1, 2, 3$ and functions $\boldsymbol{F}^k$ and $\boldsymbol{G}^k$ represent the separated functions used to approximate the unknown field, obtained in previous iterations of the PGD algorithm.

If this rank-n approximation does not give the desired accuracy, we can proceed further and look for the $(n+1)$th term, that will look like

$$u_j^{n+1}(\boldsymbol{x}, \boldsymbol{s}) = u_j^n(\boldsymbol{x}, \boldsymbol{s}) + R_j(\boldsymbol{x}) \cdot S_j(\boldsymbol{s}),$$

where $\boldsymbol{R}(\boldsymbol{x})$ and $\boldsymbol{S}(\boldsymbol{s})$ are the sought functions that improve the approximation.

By just applying standard rules of variational calculus, the test function will be

$$u_j^*(\boldsymbol{x}, \boldsymbol{s}) = R_j^*(\boldsymbol{x}) \cdot S_j(\boldsymbol{s}) + R_j(\boldsymbol{x}) \cdot S_j^*(\boldsymbol{s}).$$

To determine the new functions $\boldsymbol{R}$ and $\boldsymbol{S}$ any linearization strategy could in principle be applied. In our experience, a fixed-point algorithm in which functions $\boldsymbol{R}$ and $\boldsymbol{S}$ are determined iteratively gives very good results, even if convergence is not guaranteed, as is well known. We provide details of this strategy in the next section.

3.2.1 Computation of $\boldsymbol{S}(\boldsymbol{s})$ Assuming $\boldsymbol{R}(\boldsymbol{x})$ Is Known

In this case, the admissible variation of the displacement will be given by

$$u_j^*(\boldsymbol{x}, \boldsymbol{s}) = R_j(\boldsymbol{x}) \cdot S_j^*(\boldsymbol{s}),$$

or, equivalently, $\boldsymbol{u}^*(\boldsymbol{x}, \boldsymbol{s}) = \boldsymbol{R} \circ \boldsymbol{S}^*$, where the symbol "$\circ$" denotes the so-called entry-wise, Hadamard or Schur multiplication for vectors. Once substituted into Eq. (3.1), gives

$$\int_{\bar{\Gamma}}\int_{\Omega} \nabla_s(\boldsymbol{R} \circ \boldsymbol{S}^*) : \mathbf{C} : \nabla_s\left(\sum_{k=1}^{n} \boldsymbol{F}^k \circ \boldsymbol{G}^k + \boldsymbol{R} \circ \boldsymbol{S}\right) d\Omega d\bar{\Gamma} = \int_{\bar{\Gamma}}\int_{\Gamma_{t2}} (\boldsymbol{R} \circ \boldsymbol{S}^*) \cdot \left(\sum_{k=1}^{m} \boldsymbol{f}^k \circ \boldsymbol{g}^k\right) d\Gamma d\bar{\Gamma}, \tag{3.4}$$

or, simply

$$\int_{\bar{\Gamma}}\int_{\Omega} \nabla_s(\boldsymbol{R} \circ \boldsymbol{S}^*) : \mathbf{C} : \nabla_s(\boldsymbol{R} \circ \boldsymbol{S}) d\Omega d\bar{\Gamma}$$
$$= \int_{\bar{\Gamma}}\int_{\Gamma_{t2}} (\boldsymbol{R} \circ \boldsymbol{S}^*) \cdot \left(\sum_{k=1}^{m} \boldsymbol{f}^k \circ \boldsymbol{g}^k\right) d\Gamma d\bar{\Gamma} - \int_{\bar{\Gamma}}\int_{\Omega} \nabla_s\left(\boldsymbol{R} \circ \boldsymbol{S}^*\right) \cdot \mathcal{R}^n d\Omega d\bar{\Gamma},$$

where $\mathcal{R}^n$ represents:

$$\mathcal{R}^n = \mathbf{C} : \nabla_s \boldsymbol{u}^n.$$

Since the symmetric gradient operates on spatial variables only, we have:

$$\int_{\bar{\Gamma}}\int_{\Omega} (\nabla_s \boldsymbol{R} \circ \boldsymbol{S}^*) : \mathbf{C} : (\nabla_s \boldsymbol{R} \circ \boldsymbol{S}) d\Omega d\bar{\Gamma}$$
$$= \int_{\bar{\Gamma}}\int_{\Gamma_{t2}} (\boldsymbol{R} \circ \boldsymbol{S}^*) \cdot \left(\sum_{k=1}^{m} \boldsymbol{f}^k \circ \boldsymbol{g}^k\right) d\Gamma d\bar{\Gamma} - \int_{\bar{\Gamma}}\int_{\Omega} \left(\nabla_s \boldsymbol{R} \circ \boldsymbol{S}^*\right) \cdot \mathcal{R}^n d\Omega d\bar{\Gamma}$$

where all the terms depending on $\boldsymbol{x}$ are known. Therefore, all integrals over Ω and Γ_{t2} (support of the regularization of the initially punctual load) can be computed to obtain an equation for $\boldsymbol{S}(\boldsymbol{s})$.

3.2.2 Computation of $\boldsymbol{R}(\boldsymbol{x})$ Assuming $\boldsymbol{S}(\boldsymbol{s})$ Is Known

By proceeding in the same way, we have

$$u_j^*(\boldsymbol{x}, \boldsymbol{s}) = R_j^*(\boldsymbol{x}) \cdot S_j(\boldsymbol{s}),$$

which, once substituted into Eq. (3.1), gives

$$\int_{\bar{\Gamma}}\int_{\Omega} \nabla_s(\boldsymbol{R}^* \circ \boldsymbol{S}) : \mathbf{C} : \nabla_s\left(\sum_{k=1}^{n} \boldsymbol{F}^k \circ \boldsymbol{G}^k + \boldsymbol{R} \circ \boldsymbol{S}\right) d\Omega d\bar{\Gamma} = \int_{\bar{\Gamma}}\int_{\Gamma_{t2}} (\boldsymbol{R}^* \circ \boldsymbol{S}) \cdot \left(\sum_{k=1}^{m} \boldsymbol{f}^k \circ \boldsymbol{g}^k\right) d\Gamma d\bar{\Gamma}.$$

Conversely, all the terms depending on $\boldsymbol{s}$ (load position) can be integrated over $\bar{\Gamma}$, leading to a generalized elastic problem to compute function $\boldsymbol{R}(\boldsymbol{x})$.

3.3 Matrix Structure of the Problem

As stated before, see Eq. (3.3), the essential ingredient of PGD methods is to assume the variable (here, the displacement field) to be decomposed in the form of a finite sum of separable functions, i.e.,

$$\boldsymbol{u}(\boldsymbol{x}, \boldsymbol{s}) = \sum_{i=1}^{n} \boldsymbol{F}^i(\boldsymbol{x}) \circ \boldsymbol{G}^i(\boldsymbol{s}),$$

where a dependency on the physical position of the considered point, $\boldsymbol{x}$ and the position of the applied load, $\boldsymbol{s}$ is assumed.

In the implementation introduced in Sect. 3.4 below, we enforce functions $\boldsymbol{G}^i$ to have unity norm. By doing that, the weighting parameter α_i (see Eq. (2.2) is not computed explicitly, but assumed to multiply the $\boldsymbol{F}^i$ functions (which, therefore, do not have unity norm).

As in the previous chapter, $\boldsymbol{F}^i(\boldsymbol{x})$ and $\boldsymbol{G}^i(\boldsymbol{s})$ are approximated by employing (linear in this case) finite elements, so that, at iteration n, the i-th sum of the approximation will be given by

$$\boldsymbol{u}^h(\boldsymbol{x}, \boldsymbol{s}) = \sum_{i=1}^{n} \boldsymbol{F}^i(\boldsymbol{x}) \cdot \boldsymbol{G}^i(\boldsymbol{s}) = \sum_{i=1}^{n} \left[\boldsymbol{N}^T(\boldsymbol{x})\boldsymbol{F}^i \boldsymbol{M}^T(\boldsymbol{s})\boldsymbol{G}^i\right],$$

with $\boldsymbol{N}$ and $\boldsymbol{M}$ the vectors containing the finite element shape functions defined in each separated space and $\boldsymbol{F}^i$ and $\boldsymbol{G}^i$ the vectors of nodal values at the FE mesh for the functions $\boldsymbol{F}^i(\boldsymbol{x})$ and $\boldsymbol{G}^i(\boldsymbol{s})$, respectively. For simplicity and ease of reading, we have maintained the same notation employed in Eq. (2.8). The superscript h for the discrete, finite element approximation of a variable will no longer be employed, for clarity, if there is no risk of confusion.

The code included below solves the problem of a two-dimensional cantilever beam under a load placed at an arbitrary position along its upper face. Therefore, we assume a 2D spatial approximation on $\boldsymbol{x}$ (given by simple linear triangular elements) and 1D approximation for the $\boldsymbol{s}$ direction.

When we look for a new couple of functions in the approximation, we assume that

$$\boldsymbol{u}^{n+1}(\boldsymbol{x}, \boldsymbol{s}) = \boldsymbol{u}^n(\boldsymbol{x}, \boldsymbol{s}) + \boldsymbol{R}(\boldsymbol{x})\boldsymbol{S}(\boldsymbol{s}) = \sum_{i=1}^{n} \boldsymbol{F}^i(\boldsymbol{x})\boldsymbol{G}^i(\boldsymbol{s}) + \boldsymbol{R}(\boldsymbol{x})\boldsymbol{S}(\boldsymbol{s}), \quad (3.5)$$

while

$$\boldsymbol{u}^*(\boldsymbol{x}, \boldsymbol{s}) = \boldsymbol{R}^*(\boldsymbol{x})\boldsymbol{S}(\boldsymbol{s}) + \boldsymbol{R}(\boldsymbol{x})\boldsymbol{S}^*(\boldsymbol{s}).$$

Very often, in parametric problems (like in this example) the derivatives appearing in the weak form of the problem, given by Eq. (3.4), are acting only spatial coordinates and not on the parameters, despite that in the PGD framework they are considered as actual "extra" coordinates. Therefore,

$$\nabla_s \boldsymbol{u} = \nabla_s \left[\sum_{i=1}^{n} \boldsymbol{F}^i(\boldsymbol{x}) \cdot \boldsymbol{G}^i(s) \right] + \nabla_s \left[\boldsymbol{R}(\boldsymbol{x}) \cdot \boldsymbol{S}(s) \right] = \sum_{i=1}^{n} \left[\nabla_s \boldsymbol{F}^i(\boldsymbol{x}) \right] \cdot \boldsymbol{G}^i(s) + \left[\nabla_s \boldsymbol{R}(\boldsymbol{x}) \right] \cdot \boldsymbol{S}(s)$$
$$= \sum_{i=1}^{n} \left[\nabla_s \boldsymbol{N}^T(\boldsymbol{x}) \boldsymbol{F}^i \right] \cdot \boldsymbol{M}^T(s) \boldsymbol{G}^i + \left[\nabla_s \boldsymbol{N}^T(\boldsymbol{x}) \boldsymbol{R} \right] \cdot \boldsymbol{M}^T(s) \boldsymbol{S}.$$

The terms defined before and depending on nodal values $\boldsymbol{F}_i$ and $\boldsymbol{G}_i$ (i.e., those corresponding to the already computed $\boldsymbol{u}^n$ approximation) are known, so they should be moved to the right-hand side of the final algebraic equation.

In a similar way,

$$\nabla_s \boldsymbol{u}^* = \nabla_s \boldsymbol{N}^T(\boldsymbol{x}) \boldsymbol{R} \cdot \boldsymbol{M}^T(s) \boldsymbol{S}^* + \nabla_s \boldsymbol{N}^T(\boldsymbol{x}) \boldsymbol{R}^* \cdot \boldsymbol{M}^T(s) \boldsymbol{S}.$$

The L.H.S. of the weak form, Eq. (3.4), can be easily expressed in separated form,

$$\int_{\bar{\Gamma}} \int_{\Omega} \nabla_s \boldsymbol{u}^* : \mathbf{C} : \nabla_s \boldsymbol{R}(\boldsymbol{x}) \boldsymbol{S}(s) d\Omega d\bar{\Gamma}$$
$$= \int_{\Omega} \boldsymbol{R}^T \nabla_s \boldsymbol{N}(\boldsymbol{x}) : \mathbf{C} : \nabla_s \boldsymbol{N}^T(\boldsymbol{x}) \boldsymbol{R} d\Omega \cdot \int_{\bar{\Gamma}} \boldsymbol{S}^{*T} \boldsymbol{M}(s) \boldsymbol{M}^T(s) \boldsymbol{S} d\bar{\Gamma}$$
$$+ \int_{\Omega} \boldsymbol{R}^{*T} \nabla_s \boldsymbol{N}(\boldsymbol{x}) : \mathbf{C} : \nabla_s \boldsymbol{N}^T(\boldsymbol{x}) \boldsymbol{R} d\Omega \cdot \int_{\bar{\Gamma}} \boldsymbol{S}^T \boldsymbol{M}(s) \boldsymbol{M}^T(s) \boldsymbol{S} d\bar{\Gamma} \tag{3.6}$$

Since $\boldsymbol{R}$ and $\boldsymbol{S}$ represent vectors of nodal values for functions $\boldsymbol{R}(\boldsymbol{x})$ and $\boldsymbol{S}(s)$, respectively, they can be extracted from the integrals in Eq. (3.6), so as to give,

$$\int_{\bar{\Gamma}} \int_{\Omega} \nabla_s \boldsymbol{u}^* : \mathbf{C} : \nabla_s \boldsymbol{R}(\boldsymbol{x}) \boldsymbol{S}(s) d\Omega d\bar{\Gamma}$$
$$= \boldsymbol{R}^T \left[\int_{\Omega} \nabla_s \boldsymbol{N}(\boldsymbol{x}) : \mathbf{C} : \nabla_s \boldsymbol{N}^T(\boldsymbol{x}) d\Omega \right] \boldsymbol{R} \cdot \boldsymbol{S}^{*T} \left[\int_{\bar{\Gamma}} \boldsymbol{M}(s) \boldsymbol{M}^T(s) \right] \boldsymbol{S} d\bar{\Gamma}$$
$$+ \boldsymbol{R}^{*T} \left[\int_{\Omega} \nabla_s \boldsymbol{N}(\boldsymbol{x}) : \mathbf{C} : \nabla_s \boldsymbol{N}^T(\boldsymbol{x}) d\Omega \right] \boldsymbol{R} \cdot \boldsymbol{S}^T \left[\int_{\bar{\Gamma}} \boldsymbol{M}(s) \boldsymbol{M}^T(s) d\bar{\Gamma} \right] \boldsymbol{S}$$
$$= \boldsymbol{R}^T \boldsymbol{K1}(\boldsymbol{x}) \boldsymbol{R} \cdot \boldsymbol{S}^{*T} \boldsymbol{M2}(s) \boldsymbol{S} + \boldsymbol{R}^{*T} \boldsymbol{K1}(\boldsymbol{x}) \boldsymbol{R} \cdot \boldsymbol{S}^T \boldsymbol{M2}(s) \boldsymbol{S} \tag{3.7}$$

where $\boldsymbol{K1}(\boldsymbol{x})$ represents a sort of stiffness matrix on the spatial coordinates $\boldsymbol{x}$, while $\boldsymbol{M2}(\boldsymbol{s})$ represents a mass matrix for the 1D discretization problem in the $\boldsymbol{s}$ variable. Both of them are computed in routine `elemstiff` (see the call `[K1,M2] = elemstiff(coors)` in the main file of the code in Sect. 3.4).

We proceed similarly for the RHS term of the weak form, Eq. (3.4), and taking into account that the integrals involved in it do not depend on spatial coordinates, but only on the $\boldsymbol{s}$ coordinate, we arrive at

$$\int_{\bar{\Gamma}}\int_{\Gamma_{t2}} \boldsymbol{u}^* \cdot \boldsymbol{t} d\Gamma d\bar{\Gamma} = \int_{\bar{\Gamma}}\int_{\Gamma_{t2}} \left[\boldsymbol{R}^*(\boldsymbol{x})\boldsymbol{S}(\boldsymbol{s}) + \boldsymbol{R}(\boldsymbol{x})\boldsymbol{S}^*(\boldsymbol{s})\right] \cdot \left[\sum_{i=1}^{m} \boldsymbol{f}^i(\boldsymbol{x})\boldsymbol{g}^i(\boldsymbol{s})\right] d\Gamma d\bar{\Gamma}$$
$$= \int_{\Gamma_{t2}} \sum_{i=1}^{m} \boldsymbol{R}^*(\boldsymbol{x})\boldsymbol{f}^i(\boldsymbol{x}) \int_{\bar{\Gamma}} \boldsymbol{S}(\boldsymbol{s})\boldsymbol{g}^i(\boldsymbol{s}) d\bar{\Gamma} + \int_{\Gamma_{t2}} \sum_{i=1}^{m} \boldsymbol{R}(\boldsymbol{x})\boldsymbol{f}^i(\boldsymbol{x}) \int_{\bar{\Gamma}} \boldsymbol{S}^*(\boldsymbol{s})\boldsymbol{g}^i(\boldsymbol{s}) d\bar{\Gamma}. \tag{3.8}$$

Denoting by $\boldsymbol{FR1}$ the nodal values of the source term decomposition for the spatial variables and by $\boldsymbol{FR2}$ for the load direction, Eq. (3.8) has the form,

$$\sum_{i=1}^{m} \boldsymbol{R}^{*T} \boldsymbol{FR1} \cdot \boldsymbol{S}^T \left[\int_{\bar{\Gamma}} \boldsymbol{M}(\boldsymbol{s})\boldsymbol{M}^T(\boldsymbol{s}) d\bar{\Gamma}\right] \boldsymbol{FR2} + \sum_{i=1}^{m} \boldsymbol{R}^T \boldsymbol{FR1} \cdot \boldsymbol{S}^{*T} \left[\int_{\bar{\Gamma}} \boldsymbol{M}(\boldsymbol{s})\boldsymbol{M}^T(\boldsymbol{s}) d\bar{\Gamma}\right] \boldsymbol{FR2}. \tag{3.9}$$

As discussed in the introduction to this chapter, the problem of expressing the moving load as a finite sum of separable functions, i.e., $\boldsymbol{t}(\boldsymbol{x}, \boldsymbol{s}) \approx \sum_{i=1}^{m} \boldsymbol{f}^i(\boldsymbol{x})\boldsymbol{g}^i(\boldsymbol{s})$ is not separable. In other words, it can only be done in the discrete setting by choosing $\boldsymbol{FR1}$ to be a matrix of zeros (of size [# `dof of the whole problem` × `dof along the loaded boundary`] and whose only non-vanishing entries are, for each column, the position of nodes that can receive a load. These entries will be equal to the value of the applied force. Respectively, $\boldsymbol{FR2}$ will be an identity matrix (`eye(# dof along the loaded boundary)` in Matlab terms). Each column in $\boldsymbol{FR1}$ thus includes the corresponding degree of freedom that is loaded. Note that, therefore,

$$\sum_{i=1}^{m} \boldsymbol{f}^i(\boldsymbol{x})\boldsymbol{g}^i(\boldsymbol{s}) = \sum_{i=1}^{m} \boldsymbol{N}(\boldsymbol{x})\boldsymbol{FR1}^i \cdot \boldsymbol{M}(\boldsymbol{s})\boldsymbol{FR2}^i.$$

To this R.H.S. vector we should add the terms related to the known solution up to iteration n, $\boldsymbol{u}^n(\boldsymbol{x}, \boldsymbol{s})$, Eq. (3.5),

$$\int_{\bar{\Gamma}}\int_{\Omega} \nabla_s \boldsymbol{u}^* : \mathbf{C} : \nabla_s \boldsymbol{u}^n d\Omega d\bar{\Gamma}$$

$$= \sum_{i=1}^{n} \boldsymbol{R}^T \left[\int_{\Omega} \nabla_s \boldsymbol{N}(\boldsymbol{x}) : \mathbf{C} : \nabla_s \boldsymbol{N}^T(\boldsymbol{x}) d\Omega \right] \boldsymbol{F}^i \cdot \boldsymbol{S}^{*T} \left[\int_{\bar{\Gamma}} \boldsymbol{M}(s) \boldsymbol{M}^T(s) \right] \boldsymbol{G}^i d\bar{\Gamma} +$$

$$+ \sum_{i=1}^{n} \boldsymbol{R}^{*T} \left[\int_{\Omega} \nabla_s \boldsymbol{N}(\boldsymbol{x}) : \mathbf{C} : \nabla_s \boldsymbol{N}^T(\boldsymbol{x}) d\Omega \right] \boldsymbol{F}^i \cdot \boldsymbol{S}^{T} \left[\int_{\bar{\Gamma}} \boldsymbol{M}(s) \boldsymbol{M}^T(s) d\bar{\Gamma} \right] \boldsymbol{G}^i$$

$$= \sum_{i=1}^{n} \boldsymbol{R}^T \boldsymbol{K1}(\boldsymbol{x}) \boldsymbol{F}^i \cdot \boldsymbol{S}^{*T} \boldsymbol{M2}(s) \boldsymbol{G}^i + \sum_{i=1}^{n} \boldsymbol{R}^{*T} \boldsymbol{K1}(\boldsymbol{x}) \boldsymbol{F}^i \cdot \boldsymbol{S}^{T} \boldsymbol{M2}(s) \boldsymbol{G}^i. \tag{3.10}$$

In the Matlab code included in Sect. 3.4 we are employing the following notation: $\boldsymbol{V1}^i = \boldsymbol{FR1}^i$ and $\boldsymbol{V2}^i = \boldsymbol{M2}(s)\boldsymbol{FR2}^i$ for $i = 1, \ldots, m$, so that Eq. (3.9) is expressed in the form

$$\sum_{i=1}^{m} \boldsymbol{R}^{*T} \boldsymbol{V1} \cdot \boldsymbol{S}^T \boldsymbol{V2} + \sum_{i=1}^{m} \boldsymbol{R}^T \boldsymbol{V1} \cdot \boldsymbol{S}^{*T} \boldsymbol{V2}. \tag{3.11}$$

After some gymnastics, the weak form, Eq. (3.4), taking into account Eqs. (3.7), (3.10) and (3.11), will look in matrix form like

$$\boldsymbol{R}^T \boldsymbol{K1}(\boldsymbol{x}) \boldsymbol{R} \cdot \boldsymbol{S}^{*T} \boldsymbol{M2}(s) \boldsymbol{S} + \boldsymbol{R}^{*T} \boldsymbol{K1}(\boldsymbol{x}) \boldsymbol{R} \cdot \boldsymbol{S}^T \boldsymbol{M2}(s) \boldsymbol{S} =$$

$$= \sum_{i=1}^{m} \boldsymbol{R}^{*T} \boldsymbol{V1} \cdot \boldsymbol{S}^T \boldsymbol{V2} + \sum_{i=1}^{m} \boldsymbol{R}^T \boldsymbol{V1} \cdot \boldsymbol{S}^{*T} \boldsymbol{V2} - \sum_{i=1}^{n} \boldsymbol{R}^T \boldsymbol{K1}(\boldsymbol{x}) \boldsymbol{F}^i \cdot \boldsymbol{S}^{*T} \boldsymbol{M2}(s) \boldsymbol{G}^i$$

$$- \sum_{i=1}^{n} \boldsymbol{R}^{*T} \boldsymbol{K1}(\boldsymbol{x}) \boldsymbol{F}^i \cdot \boldsymbol{S}^T \boldsymbol{M2}(s) \boldsymbol{G}^i.$$

As mentioned before, in order to determine the new functional pair $\boldsymbol{R}$ and $\boldsymbol{S}$, any linearization strategy could be envisaged. Remember that the fact that we look for a product of functions makes the problem of enriching the approximation non-linear. In the code of Sect. 3.4, a fixed-point algorithm in which functions $\boldsymbol{R}$ and $\boldsymbol{S}$ are sought iteratively is implemented. Other strategies, like Newton-Raphson, for instance, could work equally well.

The final matrix form of the fixed-point alternating directions algorithm, in which the computation of $\boldsymbol{S}(\boldsymbol{s})$ is performed, assuming that $\boldsymbol{R}(\boldsymbol{x})$ is known, will look like

$$\boldsymbol{R}^T\boldsymbol{K1}(\boldsymbol{x})\boldsymbol{R}\cdot\boldsymbol{S}^{*T}\boldsymbol{M2}(\boldsymbol{s})\boldsymbol{S}=\sum_{i=1}^{m}\boldsymbol{R}^T\boldsymbol{V1}\cdot\boldsymbol{S}^{*T}\boldsymbol{V2}-\sum_{i=1}^{n}\boldsymbol{R}^T\boldsymbol{K1}(\boldsymbol{x})\boldsymbol{F}^i\cdot\boldsymbol{S}^{*T}\boldsymbol{M2}(\boldsymbol{s})\boldsymbol{G}^i. \tag{3.12}$$

Equivalently, when we look for $\boldsymbol{R}(\boldsymbol{x})$ assuming $\boldsymbol{S}(\boldsymbol{s})$ is known, the resulting problem will have the following matrix form,

$$\boldsymbol{R}^{*T}\boldsymbol{K1}(\boldsymbol{x})\boldsymbol{R}\cdot\boldsymbol{S}^{T}\boldsymbol{M2}(\boldsymbol{s})\boldsymbol{S}=\sum_{i=1}^{m}\boldsymbol{R}^{*T}\boldsymbol{V1}\cdot\boldsymbol{S}^{T}\boldsymbol{V2}-\sum_{i=1}^{n}\boldsymbol{R}^{*T}\boldsymbol{K1}(\boldsymbol{x})\boldsymbol{F}^i\cdot\boldsymbol{S}^{T}\boldsymbol{M2}(\boldsymbol{s})\boldsymbol{G}^i. \tag{3.13}$$

In next section the detailed Matlab code implementing this strategy is provided.

3.4 Matlab Code for the Influence Line Problem

As always, the code begins at file `main.m`, whose content is reproduced below. It solves the problem of a cantilever beam under a load placed at an arbitrary location along its top boundary. Small strains assumption is made. The code provides the solution under plane stress or plane strain conditions.

```
%
%                    PGD Code for parametrized force
%                  D. Gonzalez, I. Alfaro, E. Cueto
%                      Universidad de Zaragoza
%                        AMB-I3A Dec 2015
%
clear all; close all; clc;
%
% VARIABLES
%
global coords triangles E nu behaviour % Global variables.
E = 1000; nu = 0.3; % Material (Young Modulus and Poisson Coef)
Modulus = 1; % Force Modulus.
behaviour = 1; % Plane Stress(1), Plane Strain(2).
TOL = 1.0E-03; % Tolerance.
num_max_iter = 11; % Max. # of functional pairs for the approximation.
%
% GEOMETRY
%
X0 = 0; Xf = 3; Y0 = 0; Yf = 1.0; % The beam dimensions, [0,3]x[0,1]
tamx = 0.1; tamy = 0.1; %  mesh size along each direction
nenY = numel(Y0:tamy:Yf); % # of elements in vertical direction
[X,Y] = meshgrid(X0:tamx:Xf,Y0:tamy:Yf);
coords = [X(:),Y(:)];
force = X0:tamx:Xf; force = force'; % force positions and 1D coordinates.
triangles = delaunayTriangulation(coords(:,1),coords(:,2)); % Mesh data.
%
% ALLOCATION OF MATRICES AND VECTORS
%
F = zeros(numel(coords),1); % Nodal values of spatial function F
```

```
G = zeros(numel(force),1); % Nodal values of force function G
FR1 = zeros(numel(coords),numel(force)); % Nodal values for force (spatial term)
FR2 = eye(numel(force)); % Nodal values for force (force term)
%
% COMPUTING STIFFNESS AND MASS MATRIX FOR SPACE, ONLY MASS MATRIX FOR FORCE
%
[K1,M2] = elemstiff(force);
%
% SOURCE (FORCE) TERM IN SEPARATED FORM
%
DOFforceed = 2*nenY:2*nenY:numel(coords); % force on vertical d.o.f. on top.
for i1=1:numel(DOFforceed)
    FR1(DOFforceed(i1),i1) = -Modulus;
end
V1 = FR1; % Take into account that integration is done only in S direction
V2 = M2*FR2; % Mass matrix times nodal value of the source. R.H.S. of Eq.(3.9)
%
% BOUNDARY CONDITIONS
%
CC = 1:2*(nenY); % Left side of the beam fixed.
%
% ENRICHMENT OF THE APPROXIMATION, LOOKING FOR R AND S
%
num_iter = 0; iter = zeros(1); Aprt = 0; Error_iter = 1.0;
while Error_iter>TOL && num_iter<num_max_iter
    num_iter = num_iter + 1;
    S0 = rand(numel(force),1); % Initial guess for S.
    %
    % ENRICHMENT STEP
    %
    [R,S,iter(num_iter)] = enrichment(K1,M2,V1,V2,S0,F,G,num_iter,TOL,CC);
    F(:,num_iter) = R; G(:,num_iter) = S; % R and S are valid, new summand.
    %
    % STOPPING CRITERION
    %
    Error_iter = norm(F(:,num_iter)*G(:,num_iter)');
    Aprt = max(Aprt,sqrt(Error_iter));
    Error_iter = sqrt(Error_iter)/Aprt;
    fprintf(1,'%dst summand in %d iterations with a weight of %f\n',...
        num_iter,iter(num_iter),Error_iter);
end
num_iter = num_iter - 1; % The last sum was negligible, we discard it.
fprintf(1,'PGD off-line Process exited normally\n\n');
save('WorkSpacePGD_Parametricedforce.mat');
%
% POST-PROCESSING
%
fprintf(1,'Please select force position');
fprintf(1,'on the figure or pick out of the beam to exit');
h1 = figure(1); triplot(triangles);
axis equal;
[Cx,Cy] = ginput(1); % Waiting for a mouse click on the figure.
lim = 0.2/(Xf-X0); % Establishes an exit zone on the figure.
while X0-lim<=Cx && Cx<=Xf+lim && Y0-lim<=Cy && Cy<=Yf+lim
    h1 = figure(1); triplot(triangles);
    axis equal;
    Posforce = find(force<Cx,1,'last'); % Look for the closest loaded node.
    %
    % EVALUATING THE SOLUTION CHOOSING THE SELECTED NODE IN G VECTOR
    %
    desp = zeros(numel(coords),1);
    for i1=1:num_iter
        desp = desp + F(:,i1).*G(Posforce,i1); % Obtain the solution
    end
    %
    % PLOTTING THE SOLUTION
    %
    cdx = coords(:,1) + desp(1:2:end); % New X coordinates.
    cdy = coords(:,2) + desp(2:2:end); % New Y coordinates.
```

```
    trisurf(triangles.ConnectivityList,cdx,cdy,desp(2:2:end));
    title('Vertical_Displacement'); view(2); colorbar;
    figure(1); axis equal; [Cx,Cy] = ginput(1); % Wait for a new force.

end
fprintf(1,'\n\n##########_End_of_simulation_##########\n\n');
```

As in Chap. 2, what we call stiffness and mass matrices are computed in function `elemstiff.m`:

```
function [K1,M2] = elemstiff(coor2)
% function [K1,M2] = elemstiff(coor2)
% For space compute stifness matrix, for load parameter compute mass matrix
% Universidad de Zaragoza - 2015

%
% SPACE MATRICES
%
[K1] = fem2D; % Standard 2D FEM code for Triangular Elements, computing
%
% LOAD MATRICES: 1D PARAMETRIC PROBLEM
%
sg = [-1.0/sqrt(3.0) 1.0/sqrt(3.0)]; wg = ones(2,1); % Gauss points
npg = numel(sg); nen2 = numel(coor2); M2 = zeros(nen2);
X2 = coor2(1:nen2-1)'; Y2 = coor2(2:nen2)'; % Coordinates of elements
L2 = Y2 - X2; % Longitude of each  element for parametriced variable
for i1=1:nen2-1
    c2 = zeros(1,npg); N2 = zeros(nen2,npg);
    c2(1,:) = 0.5.*(1.0-sg).*X2(i1) + 0.5.*(1.0+sg).*Y2(i1);
    N2(i1+1,:) = (c2(1,:)-X2(i1))./L2(i1);
    N2(i1,:) = (Y2(i1)-c2(1,:))./L2(i1);
    for j1=1:npg
        M2 = M2 + N2(:,j1)*N2(:,j1)'*0.5.*wg(j1).*L2(i1); %NûN
    end
end
return
```

The enrichment procedure, i.e., the computation of a new functional pair $\boldsymbol{R}$, $\boldsymbol{S}$, is detailed in function `enrichment.m`:

```
function  [R,S,iter] = enrichment(K1,M2,V1,V2,S0,F,G,num_iter,TOL,CC)
% function  [R,S,iter] = enrichment(K1,M2,V1,V2,S0,F,G,num_iter,TOL,CC)
% Computes a new sumand by fixed-point algorithm using PGD
% Universidad de Zaragoza - 2015
R = zeros(size(F,1),1); R0 = R; % Initial value R to compare in first loop.
h = size(V2,2); % Number functions of the source
ExitFlag = 1;
iter = 0;
mxit = 25; % #ã of possible iterations for the fixde point algorithm.
Free = setdiff(1:numel(F(:,1)),CC);
%
% FIXED POINT ALGORITHM
%
while ExitFlag>TOL
    %
    % LOOKING FOR R, KNOWNING S
    %
    matrixR = K1*(S0'*M2*S0);
    sourceR = zeros(size(F,1),1);
    for k1=1:h
        sourceR = sourceR + V1(:,k1)*(S0'*V2(:,k1));
    end
    for i1=1:num_iter-1
        sourceR = sourceR - K1*F(:,i1)*(S0'*M2*G(:,i1));
    end
```

```
    %
    % SOLVE R
    %
    R(Free) = matrixR(Free,Free)\sourceR(Free);
    %
    % LOOKING FOR S, KNOWNING R
    %
    matrixS = (R'*K1*R)*M2;
    sourceS = zeros(size(G,1),1);
    for k1=1:h
        sourceS = sourceS + V2(:,k1)*(R'*V1(:,k1));
    end
    for i1=1:num_iter-1
        sourceS = sourceS - R'*K1*F(:,i1)*(M2*G(:,i1));
    end
    %
    % SOLVE S
    %
    S = matrixS\sourceS;
    S = S./norm(S); % We normalize S. R takes care of alpha constant.
    %
    % COMPUTING STOP CRITERIA
    %
    error = max(abs(sum(R0-R)),abs(sum(S0-S))); R0 = R; S0 = S;
    iter = iter + 1;
    if iter>mxit !! abs(error)<TOL,
        return
    end
end
return
```

The code makes use of a traditional, two-dimensional FEM code, whose structure is reproduced below. In fact, it returns the stiffness matrix $\boldsymbol{K}$ typical of these FEM programs.

```
function [K] = fem2D
% function [K] = fem2D
% A 2D FEM code for linear triangles. Return Stifness matrix
% Universidad de Zaragoza - 2015
global coords triangles E nu behaviour
dof = 2; % Degree of freedom per node
numNodes = size(coords,1); numTriang = size(triangles,1);
%
% ALLOCATE MEMORY
%
K = zeros(dof*numNodes);
%
% MATERIAL AND BEHAVIOUR
%
G = E/2/(1+nu);
if behaviour==2 % Plane Strain
    E1 = E*(1-nu)/(1+nu)/(1-2*nu);
    E2 = E*nu/(1+nu)/(1-2*nu);
elseif behaviour==1 % Plane Stress
    E1 = E/(1-nu^2);
    E2 = E*nu/(1-nu^2);
end
D = [E1 E2 0;E2 E1 0;0 0 G]; % Behaviour matrix
% Integration points: 3 Hammer Points
sg(1) = 1.0/6.0; sg(2) = 1.0/6.0; sg(3) = 2.0/3.0;
sg(4) = 1.0/6.0; sg(5) = 1.0/6.0; sg(6) = 2.0/3.0;
wg(1) = 1.0/6.0; wg(2) = 1.0/6.0; wg(3) = 1.0/6.0;
nph = numel(wg);
%
% ELEMENT LOOP
```

```
%
for j=1:numTriang
    tri = triangles.ConnectivityList(j,:); % Connectivity of each Element
    vertices = coords(tri,:); % Coordinates of the nodes
    Ind = [2*(tri-1)+1; 2*tri]; Ind = Ind(:);
    %
    % JACOBIAN
    %
    a = vertices(2,1)-vertices(1,1); b = vertices(3,1)-vertices(1,1);
    c = vertices(2,2)-vertices(1,2); d = vertices(3,2)-vertices(1,2);
    jcob = a*d - b*c;
    a1 = vertices(2,1)*vertices(3,2) - vertices(2,2)*vertices(3,1);
    a2 = vertices(3,1)*vertices(1,2) - vertices(3,2)*vertices(1,1);
    a3 = vertices(1,1)*vertices(2,2) - vertices(1,2)*vertices(2,1);
    b1 = vertices(2,2) - vertices(3,2); b2 = vertices(3,2) - vertices(1,2);
    b3 = vertices(1,2) - vertices(2,2);
    c1 = vertices(3,1) - vertices(2,1); c2 = vertices(1,1) - vertices(3,1);
    c3 = vertices(2,1) - vertices(1,1);
    %
    % INTEGRATION POINTS LOOP
    %
    chiG = 0.0; etaG = 0.0;
    for j1=1:nph
        chi = sg(2*(j1-1)+1); eta = sg(2*j1);
        %
        % GLOBAL GEOMETRICAL APPROXIMATION
        %
        SHPa(3) = eta; SHPa(2) = chi; SHPa(1) = 1.-chi -eta;
        for k1=1:3
            chiG = chiG + SHPa(k1)*vertices(k1,1);
            etaG = etaG + SHPa(k1)*vertices(k1,2);
        end
        %
        % COMPUTE SHAPE FUNCTIONS AND THEIR DERIVATIVES
        %
        SHP(1) = (a1+b1*chiG+c1*etaG)/jcob; dSHPx(1) = b1; dSHPy(1) = c1;
        SHP(2) = (a2+b2*chiG+c2*etaG)/jcob; dSHPx(2) = b2; dSHPy(2) = c2;
        SHP(3) = (a3+b3*chiG+c3*etaG)/jcob; dSHPx(3) = b3; dSHPy(3) = c3;
        %
        % N AND B MATRIX
        %
        B = [dSHPx(1) 0 dSHPx(2) 0 dSHPx(3) 0; ...
            0 dSHPy(1) 0 dSHPy(2) 0 dSHPy(3); ...
            dSHPy(1) dSHPx(1) dSHPy(2) dSHPx(2) dSHPy(3) dSHPx(3)];
        %
        % STIFNESS MATRIX
        %
        K(Ind,Ind) = K(Ind,Ind) + B'*D*B/jcob*wg(j1);
    end
end
return
```

Once executed, the code allows the user to choose interactively with the mouse the point in which the load is applied. It is implemented in an off-line/on-line approach, such that the modes (functional pairs) approximating the solution are first computed and (eventually) stored in memory. Then, in the on-line phase, the user can interactively play with the position of the load and see in real time the deformed configuration of the solid.

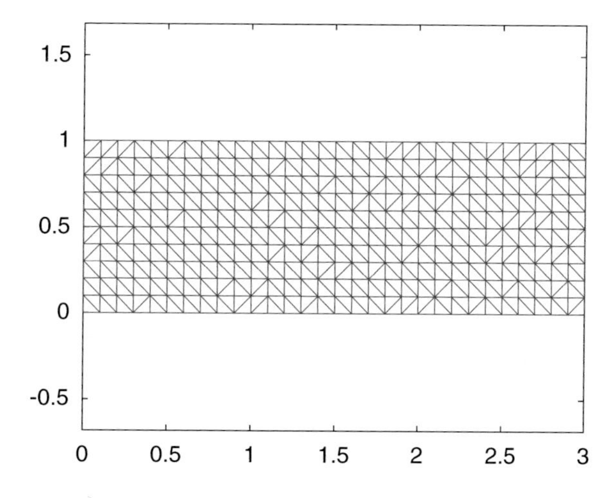

Fig. 3.4 Mesh for the moving load problem

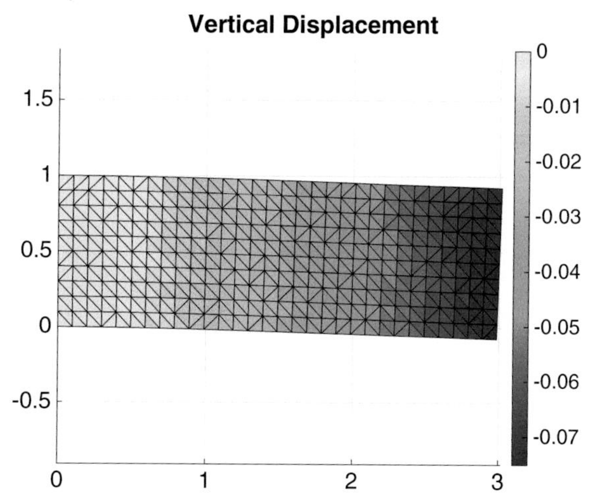

Fig. 3.5 Once a point along the upper side of the beam has been chosen, the problem depicts the deformed configuration of the beam

At a first instance, the mesh of the problem is shown, see Fig. 3.4. By clicking on it, the user can choose the particular placement of the load. The program gives immediately the deformed configuration of the beam, see Fig. 3.5.

Note that by simply typing on the Matlab command line the instruction `trisurf(triangles.ConnectivityList,coords(:,1),coords(:,2), F(1:2:end,1))`, the first spatial mode of the solution, namely $\boldsymbol{F}^1(\boldsymbol{x})$ is represented, see Fig. 3.6. Equivalently, by typing `plot(G(:,1))` the load modes $\boldsymbol{G}^i(\boldsymbol{s})$ can be plotted, see Fig. 3.7. Notice the increasing frequency content of the modes in the load variable.

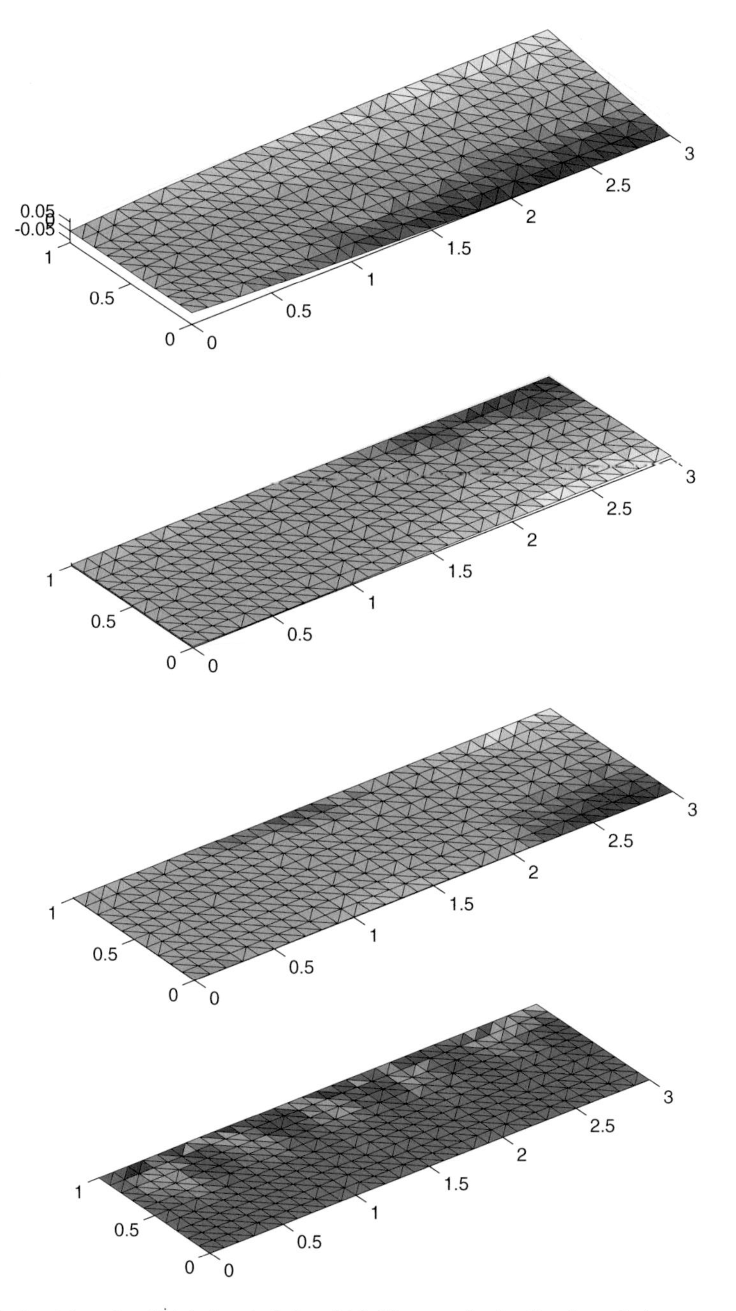

Fig. 3.6 Spatial modes $\boldsymbol{F}^i(\boldsymbol{x})$, $i = 1, 2, 3$ and 11. The magnitude of each mode is represented in the *vertical axis*

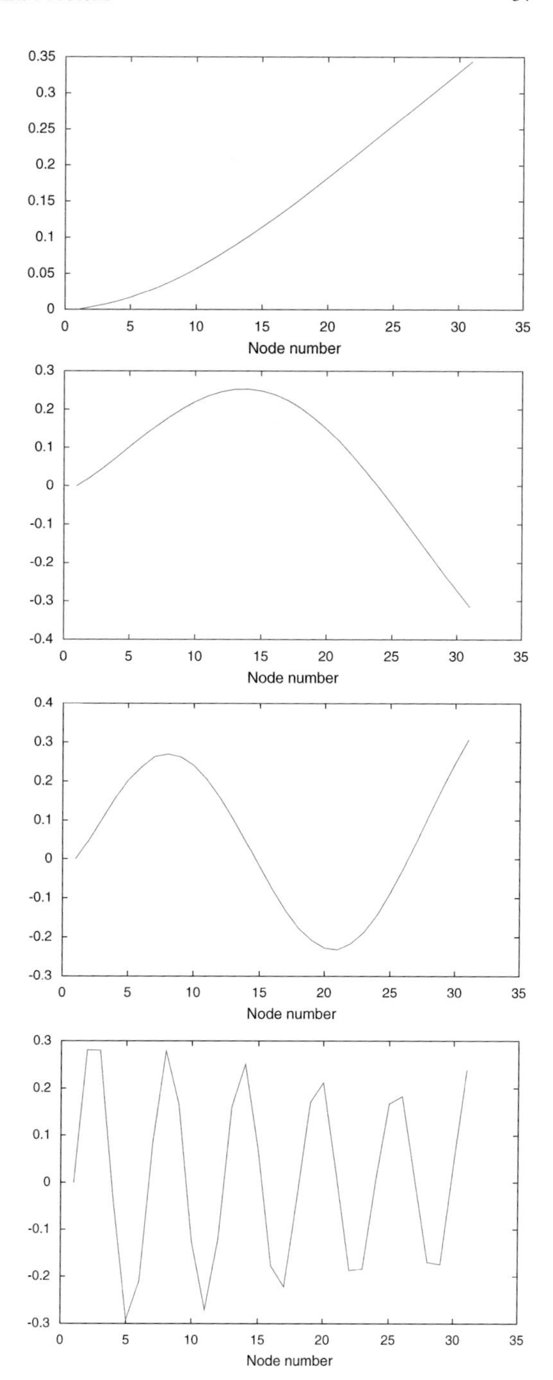

Fig. 3.7 Load modes $\boldsymbol{G}^i(s)$, $i = 1, 2, 3$ and 11. Node labels refer to the relative position along the upper boundary of the beam

Fig. 3.8 A virtual surgery simulator based on the same algorithm explained in this chapter

This algorithm has revealed to be very powerful. In fact, it is essentially the same employed to construct our virtual surgery simulator, see Fig. 3.8, able to provide response feedback in the order of kHz, thus amenable to be employed in haptic environments [56, 57].

Chapter 4
PGD for Non-linear Problems

Classical mathematics concentrated on linear equations for a sound pragmatic reason: it could not solve anything else.
—Ian Stewart

Abstract This chapter develops the application of PGD methods for linear elasticity problems. The only difficulty relies in the vectorial character of the unknown, the displacement field.

Non-linear problems continue to be the bottleneck of model order reduction techniques. This is so since most of the reduction in computational complexity is lost if it continues to be necessary to reconstruct the tangent stiffness matrix of the full order problem. Some years ago, reduced order models were in fact linear [54] precisely to avoid this pitfall. Since then, several techniques have been developed in the field, while it continues to be one of the most active areas of research within the model order reduction community. Without being exhaustive, one can cite very different approaches to model order reduction of non-linear problems. One of the earliest is perhaps the Large Time Increment (LaTIn) method [32, 44, 45]. In the context of PGD techniques, one of the earliest attempts to solve this kind of problems was to couple with Asymptotic Numerical (ANM) Methods [1, 20, 31, 62], giving rise to a sort of coupled PGD-ANM method [55, 56]. This method works well for smooth non-linearities such as those appearing in hyperelasticity problems.

For more general types of non-linearities, the empirical interpolation method (EIM) by Maday and coworkers remains to be amongst the most popular techniques [14]. It consists, roughly, in determining the best possible interpolation points so as to interpolate the non-linear terms of the equation. Very closely related, the discrete empirical interpolation method (DEIM) [21] has gained a lot of popularity in recent times, like the so-called "best points" interpolation method [53] and related approaches [12]. Other techniques that have been employed to some extent in

E. Cueto et al., *Proper Generalized Decompositions*, SpringerBriefs in Applied Sciences and Technology, DOI 10.1007/978-3-319-29994-5_4

trying to overcome the difficulties posed by non-linearities include proper orthogonal decomposition with interpolation, PODI [50] or the reduced-order interpoaltion technique developed by Farhat and coworkers based upon the concept of Grasmann manifolds, [12].

Despite all these attempts, one can say that no definitive answer has been provided for the problem of reducing a non-linear problem. The difficulties come from a variety of sources and no single technique gives a solution free of drawbacks. In this chapter we are going to explain a very simple one, based on an explicit linearization of the problem. To that end, we will take non-linear hyperelasticity as a model problem. In essence, this approach is the same introduced in some of our previous works, see [56].

4.1 Hyperelasticity

The simplest non-linear hyperelastic model is the so-called Kirchhoff-Saint Venant model. It consists roughly in applying the traditional linear elasticity constitutive equations upon non linear strain measures. It is well-known, however, that it leads to instabilities in compression, which makes it to be very rarely used in practice. However, it includes most of the difficulties of more complex models, such as neo-Hookean models [56].

In essence, the Kirchhoff-Saint Venant model (linear elastic material under non-linear strain measures) can be derived from a potential of the type

$$\Psi = \frac{\lambda}{2}(\mathrm{tr}(\boldsymbol{E}))^2 + \mu \boldsymbol{E} : \boldsymbol{E},$$

where $\boldsymbol{E}$ stands for the Green-Lagrange strain tensor, and λ and μ are the Lame's coefficients. The Green-Lagrange strain tensor is indeed formed by a linear and a non-linear part,

$$\boldsymbol{E} = \frac{1}{2}(\boldsymbol{F}^T \boldsymbol{F} - \boldsymbol{I}) = \nabla_s \boldsymbol{u} + \frac{1}{2}(\nabla^T \boldsymbol{u} \cdot \nabla \boldsymbol{u}), \tag{4.1}$$

where $\boldsymbol{F}$ represents the deformation gradient. In turn, the second Piola-Kirchhoff stress tensor can be obtained after the Green-Lagrange strain tensor by applying the appropriate fourth-order constitutive tensor:

$$\boldsymbol{S} = \frac{\partial \Psi(\boldsymbol{E})}{\partial \boldsymbol{E}} = \mathbf{C} : \boldsymbol{E}.$$

The weak form of the problem can be obtained after noticing that $\boldsymbol{E}$ and $\boldsymbol{S}$ are conjugate magnitudes, i.e., their product gives the virtual work done by the material undergoing deformation:

find the displacement $\boldsymbol{u} \in \mathcal{H}^1$ *such that for all* $\boldsymbol{u}^* \in \mathcal{H}^1_0$:

$$\int_{\bar{\Gamma}} \int_{\Omega} \boldsymbol{E}^* : \mathbf{C} : \boldsymbol{E} d\Omega d\bar{\Gamma} = \int_{\bar{\Gamma}} \int_{\Gamma_{t2}} \triangle \boldsymbol{u}^* \cdot \boldsymbol{t} d\Gamma d\bar{\Gamma}, \tag{4.2}$$

where $\Gamma = \Gamma_u \cup \Gamma_t$ represents the boundary of the solid, divided into essential and natural regions, and where $\Gamma_t = \Gamma_{t1} \cup \Gamma_{t2}$, i.e., regions of homogeneous and non-homogeneous, respectively, natural boundary conditions. $\bar{\Gamma}$ represents the portion of Γ_{t2} where the (parametric) load can be applied. Note that we consider not only a non-linear problem (arising from the non-linear strain measures), but consider the parametric character of the formulation presented in Chap. 3. In other words, we are considering the position of the applied load as a parameter in the formulation and hence the doubly-weak form of the problem in Eq. (4.2).

As usual in the finite element community, we consider a force-driven problem, and therefore we apply the load in small increments. The unknown will be therefore $\triangle \boldsymbol{u}$, thus $\boldsymbol{u}^{t+\Delta t} = \boldsymbol{u}^t + \triangle \boldsymbol{u}$ and

$$\boldsymbol{E}^{t+\Delta t} = \nabla_s \left(\boldsymbol{u}^t + \triangle \boldsymbol{u} \right) + \frac{1}{2} \left(\nabla^T (\boldsymbol{u}^t + \triangle \boldsymbol{u}) \cdot \nabla (\boldsymbol{u}^t + \triangle \boldsymbol{u}) \right). \tag{4.3}$$

Similarly, the admissible variation of strain reads

$$\begin{aligned} \boldsymbol{E}^* &= \nabla_s(\triangle \boldsymbol{u}^*) + \frac{1}{2}(\nabla^T(\triangle \boldsymbol{u}^*)) \cdot \nabla(\boldsymbol{u}^t + \triangle \boldsymbol{u}) + \frac{1}{2}\nabla^T(\boldsymbol{u}^t + \triangle \boldsymbol{u}) \cdot \nabla(\triangle \boldsymbol{u}^*) \\ &= \nabla_s(\triangle \boldsymbol{u}^*) + \nabla^T(\triangle \boldsymbol{u}^*) \cdot \nabla(\boldsymbol{u}^t + \triangle \boldsymbol{u}) \end{aligned} \tag{4.4}$$

By introducing Eqs. (4.3) and (4.4) into the weak form of the problem, Eq. (4.2), we arrive at

$$\begin{aligned} \int_{\bar{\Gamma}} \int_{\Omega(t)} \boldsymbol{E}^* : \mathbf{C} : \boldsymbol{E} d\Omega d\bar{\Gamma} = \int_{\bar{\Gamma}} \int_{\Omega(t)} &\left(\nabla_s(\triangle \boldsymbol{u}^*) + \nabla(\triangle \boldsymbol{u}^*) \cdot \nabla^T(\boldsymbol{u}^t + \triangle \boldsymbol{u}) \right) : \mathbf{C} \\ &: \left(\nabla_s \left(\boldsymbol{u}^t + \triangle \boldsymbol{u} \right) + \frac{1}{2} \left(\nabla(\boldsymbol{u}^t + \triangle \boldsymbol{u}) \cdot \nabla^T(\boldsymbol{u}^t + \triangle \boldsymbol{u}) \right) \right) d\Omega d\bar{\Gamma}. \end{aligned} \tag{4.5}$$

The simplest linearization of Eq. (4.5) consists of keeping in the formulation only constant terms and those linear in $\triangle \boldsymbol{u}$. For the left-hand side term of Eq. (4.2), this gives rise to ten terms, whose precise form is detailed here:

$$\int_{\bar{\Gamma}} \int_{\Omega} \boldsymbol{E}^* : \mathbf{C} : \boldsymbol{E} d\Omega d\bar{\Gamma} \approx T1 + T2 + T3 + T4 + T5 + T6 + T7 + T8 + T9 + T10, \tag{4.6}$$

where,

$$
\begin{aligned}
T1 &= \int_{\bar{\Gamma}}\int_{\Omega} \nabla_s(\Delta \boldsymbol{u}^*) : \mathbf{C} : \nabla_s \boldsymbol{u}^t d\Omega d\bar{\Gamma}, \\
T2 &= \int_{\bar{\Gamma}}\int_{\Omega} \nabla_s(\Delta \boldsymbol{u}^*) : \mathbf{C} : \nabla_s(\Delta \boldsymbol{u}) d\Omega d\bar{\Gamma}, \\
T3 &= \int_{\bar{\Gamma}}\int_{\Omega} \nabla_s(\Delta \boldsymbol{u}^*) : \mathbf{C} : \frac{1}{2}\nabla^T \boldsymbol{u}^t \cdot \nabla \boldsymbol{u}^t d\Omega d\bar{\Gamma}, \\
T4 &= \int_{\bar{\Gamma}}\int_{\Omega} \nabla_s(\Delta \boldsymbol{u}^*) : \mathbf{C} : \nabla^T \boldsymbol{u}^t \cdot \nabla(\Delta \boldsymbol{u}) d\Omega d\bar{\Gamma}, \\
T5 &= \int_{\bar{\Gamma}}\int_{\Omega} \nabla^T(\Delta \boldsymbol{u}^*) \cdot \nabla \boldsymbol{u}^t : \mathbf{C} : \nabla_s \boldsymbol{u}^t d\Omega d\bar{\Gamma}, \\
T6 &= \int_{\bar{\Gamma}}\int_{\Omega} \nabla^T(\Delta \boldsymbol{u}^*) \cdot \nabla \boldsymbol{u}^t : \mathbf{C} : \nabla_s(\Delta \boldsymbol{u}) d\Omega d\bar{\Gamma}, \\
T7 &= \int_{\bar{\Gamma}}\int_{\Omega} \nabla^T(\Delta \boldsymbol{u}^*) \cdot \nabla \boldsymbol{u}^t : \mathbf{C} : \frac{1}{2}\nabla^T \boldsymbol{u}^t \cdot \nabla \boldsymbol{u}^t d\Omega d\bar{\Gamma}, \\
T8 &= \int_{\bar{\Gamma}}\int_{\Omega} \nabla^T(\Delta \boldsymbol{u}^*) \cdot \nabla \boldsymbol{u}^t : \mathbf{C} : \nabla^T \boldsymbol{u}^t \cdot \nabla(\Delta \boldsymbol{u}) d\Omega d\bar{\Gamma}, \\
T9 &= \int_{\bar{\Gamma}}\int_{\Omega} \nabla^T(\Delta \boldsymbol{u}^*) \cdot \nabla(\Delta \boldsymbol{u}) : \mathbf{C} : \nabla_s \boldsymbol{u}^t d\Omega d\bar{\Gamma}, \\
T10 &= \int_{\bar{\Gamma}}\int_{\Omega} \nabla^T(\Delta \boldsymbol{u}^*) \cdot \nabla(\Delta \boldsymbol{u}) : \mathbf{C} : \frac{1}{2}\nabla^T \boldsymbol{u}^t \cdot \nabla \boldsymbol{u}^t d\Omega d\bar{\Gamma}.
\end{aligned}
$$

The standard forward-Euler scheme can also be applied to this problem. However, for the tests we have performed, instabilities in compression appeared, a typical characteristic of Kirchhoff-Saint Venant models [17], as mentioned before. We have not found any instability by applying the before introduced algorithm, even if they can appear as a consequence of the characteristics of Kirchhoff-Saint Venant materials.

It is important ot note that terms $T1 + T3 + T5 + T7$ give precisely $\int_{\bar{\Gamma}}\int_{\Gamma_{t2}} \Delta \boldsymbol{u}^* \cdot \boldsymbol{t}^t d\Gamma d\bar{\Gamma}$, i.e., equilibrium at time u^t, which is assumed to have converged, and therefore to be fulfilled. What remains to be computed, in fact, is the equilibrium equation for the subsequent time step,

$$
T2 + T4 + T6 + T8 + T9 + T10 = \int_{\bar{\Gamma}}\int_{\Gamma_{t2}} \Delta \boldsymbol{u}^* \cdot \Delta \boldsymbol{t} d\Gamma d\bar{\Gamma}.
$$

It is time now to invoke the separation of variables typical of PGD strategies, so as to give, at iteration n,

$$\Delta u_i^n(\boldsymbol{x}, \boldsymbol{s}) = \sum_{j=1}^{n} F_i^j(\boldsymbol{x}) \cdot G_i^j(\boldsymbol{s}).$$

Conversely, at iteration $n+1$ of the greedy algorithm we seek for two new enrichment functions, $R(\boldsymbol{x})$ and $S(\boldsymbol{s})$,

$$\Delta u_i^{n+1}(\boldsymbol{x}, \boldsymbol{s}) = \Delta u_i^n(\boldsymbol{x}, \boldsymbol{s}) + R_i(\boldsymbol{x}) \cdot S_i(\boldsymbol{s}).$$

By computing the admissible variation of the displacement, we obtain

$$u_i^*(\boldsymbol{x}, \boldsymbol{s}) = R_i^*(\boldsymbol{x}) \cdot S(\boldsymbol{s}) + R_i(\boldsymbol{x}) \cdot S^*(\boldsymbol{s}).$$

By summing up all the displacement increments along the time history, we obtain the last converged displacement.[1] We assume that this has been done with the help of some `nprev` functional couples,

$$u_i^t(\boldsymbol{x}, \boldsymbol{s}) \approx \sum_{j=1}^{\mathtt{nprev}} F_i^j(\boldsymbol{x}) \cdot G^j(\boldsymbol{s}).$$

It is important to note that, for the sake of simplicity in the exposition, we have chosen, as in Chap. 3 a load of unity modulus, acting always in the vertical direction. No follower loads are considered. Therefore, the reader must note that $\boldsymbol{F}^j$ and $\boldsymbol{R}$ are indeed vectors, defined in the whole geometry of the solid, Ω while G^j and S are scalars defined on $\bar{\Gamma}$.

In the following Section details of the Matlab implementation are given.

4.2 Matrix Structure of the Problem

In what follows a three-dimensional problem is studied under the just presented rationale. Its two-dimensional counterpart would be essentially identical, with the sole exception of changing the finite element shape functions.

As in the previous applications of PGD methods, the seek for an enrichment functional couple is often accomplished in the form of an alternating directions

[1] Note that we employ the word *converged* since the PGD algorithm implies the solution of a non-linear problem by an alternating directions strategy. Keep in mind that the proposed linearization for the global problem is indeed explicit, so there is no need to *converge* in a Newton-Raphson sense.

algorithm in which we look alternatively for functions $\boldsymbol{R}(\boldsymbol{x})$ and $S(\boldsymbol{s})$. In the just presented explicit linearization the essential variable of the problem is the current increment of displacement, Δu_i, which is in turn separated as

$$\Delta u_i(\boldsymbol{x}, \boldsymbol{s}) = \sum_{j=1}^{n} F_i^j(\boldsymbol{x}) \cdot G^j(\boldsymbol{s}) + R_i(\boldsymbol{x}) \cdot S(\boldsymbol{s}). \tag{4.7}$$

By just applying classical variational methods, we obtain the form of an admissible variation of the displacement,

$$u_i^*(\boldsymbol{x}, \boldsymbol{s}) = R_i^*(\boldsymbol{x}) \cdot S(\boldsymbol{s}) + R_i(\boldsymbol{x}) \cdot S^*(\boldsymbol{s}),$$

where the subscript i refers to the i-th component of the displacement (a three-dimensional vector, as it is well known).

If we assume iteratively that first $S(\boldsymbol{s})$ is known, then

$$u_i^*(\boldsymbol{x}, \boldsymbol{s}) = R_i^*(\boldsymbol{x}) \cdot S(\boldsymbol{s}),$$

while, if $\boldsymbol{R}(\boldsymbol{x})$ is assumed to be known, then,

$$u_i^*(\boldsymbol{x}, \boldsymbol{s}) = R_i(\boldsymbol{x}) \cdot S^*(\boldsymbol{s}).$$

We detail here the case in which $S(\boldsymbol{s})$ is assumed to be known and therefore we look for $\boldsymbol{R}(\boldsymbol{x})$, the case in which $\boldsymbol{R}(\boldsymbol{x})$ is known being straightforward. In what follows, every term in Eq. (4.6) is detailed in matrix form.

4.2.1 Matrix Form of the Term $T2$

The term $T2$ in Eq. (4.6) can in turn be decomposed into

$$T2 = \int_{\bar{\Gamma}}\int_{\Omega} \nabla_s(\boldsymbol{R}^*(\boldsymbol{x}) \cdot S(\boldsymbol{s})) : \mathbf{C} : \nabla_s \left(\sum_{j=1}^{n} \boldsymbol{F}^j(\boldsymbol{x}) \cdot G^j(\boldsymbol{s}) + \boldsymbol{R}(\boldsymbol{x}) \cdot S(\boldsymbol{s}) \right) d\Omega d\bar{\Gamma}$$
$$= T2^K + T2^F,$$

a term in the unknown vector plus a term composed by known vectors (i.e., related to functions $\boldsymbol{F}^j$ and S), that therefore will go to the "force" vector (right-hand side of the algebraic equation). The form of these terms is the following:

$$T2^K = \int_{\bar{\Gamma}} \int_{\Omega} \nabla_s(\boldsymbol{R}^*(\boldsymbol{x})) \cdot S(\boldsymbol{s}) : \mathbf{C} : \nabla_s(\boldsymbol{R}(\boldsymbol{x})) \cdot S(\boldsymbol{s}) d\Omega d\bar{\Gamma}$$
$$= \int_{\bar{\Gamma}} S(\boldsymbol{s}) \cdot S(\boldsymbol{s}) d\bar{\Gamma} \int_{\Omega} \nabla_s(\boldsymbol{R}^*(\boldsymbol{x})) : \mathbf{C} : \nabla_s(\boldsymbol{R}(\boldsymbol{x})) d\Omega,$$

and, in turn,

$$T2^F = \sum_{j=1}^{n} \int_{\bar{\Gamma}} \int_{\Omega} \nabla_s(\boldsymbol{R}^*(\boldsymbol{x})) \cdot S(\boldsymbol{s}) : \mathbf{C} : \nabla_s(\boldsymbol{F}^j(\boldsymbol{x})) \cdot G^j(\boldsymbol{s}) d\Omega d\bar{\Gamma}$$
$$= \sum_{j=1}^{n} \int_{\bar{\Gamma}} S(\boldsymbol{s}) \cdot G^j(\boldsymbol{s}) d\bar{\Gamma} \int_{\Omega} \nabla_s(\boldsymbol{R}^*(\boldsymbol{x})) : \mathbf{C} : \nabla_s(\boldsymbol{F}^j(\boldsymbol{x})) d\Omega.$$

To approximate the vectorial unknown $\boldsymbol{R}(\boldsymbol{x})$ we employ linear finite element shape functions, stored in a matrix $\boldsymbol{N}(\boldsymbol{x})$, whose derivatives are stored, as in most classical finite element books, in a matrix $\boldsymbol{B}$. Conversely, for the scalar unknown $S(\boldsymbol{s})$ and related functions, we have

$$S(\boldsymbol{s}) = \boldsymbol{M}\boldsymbol{s}, \qquad G^j(\boldsymbol{s}) = \boldsymbol{M}\boldsymbol{G}^j, \qquad \nabla_s(\boldsymbol{R}(\boldsymbol{x})) = \boldsymbol{B}\boldsymbol{R},$$
$$\nabla_s(\boldsymbol{F}^j(\boldsymbol{x})) = \boldsymbol{B}\boldsymbol{F}^j, \qquad \nabla_s(\boldsymbol{R}^*(\boldsymbol{x})) = \boldsymbol{B}\boldsymbol{R}^*,$$

with $\boldsymbol{M}$ the shape function vector employed to approximate functions S and G^j.

Note that the vector $\boldsymbol{R}$ refers here to the vector of nodal values for the function $\boldsymbol{R}(\boldsymbol{x})$, i.e., it has three components per node:

$$\boldsymbol{R}^T = [R_x^1, R_y^1, R_z^1, R_x^2, R_y^2, R_z^2, \ldots, R_z^N],$$

with N the number of nodes in the mesh.

Both terms for $T2$ will finally look like

$$T2^K = \left(\boldsymbol{s}^T \left[\int_{\bar{\Gamma}} \boldsymbol{M}^T \boldsymbol{M} d\bar{\Gamma}\right] \boldsymbol{s}\right) \left(\boldsymbol{R}^{*T} \left[\int_{\Omega} \boldsymbol{B}^T \mathbf{C} \boldsymbol{B} d\Omega\right] \boldsymbol{R}\right) = \left(\boldsymbol{s}^T \boldsymbol{s}_1 \boldsymbol{s}\right) \left(\boldsymbol{R}^{*T} \boldsymbol{r}_1 \boldsymbol{R}\right),$$

with $\boldsymbol{R}$ the vector of nodal unknowns, and

$$T2^F = \sum_{j=1}^{n} \left(\boldsymbol{s}^T \boldsymbol{s}_1 \boldsymbol{G}^j\right) \left(\boldsymbol{R}^{*T} \boldsymbol{r}_1 \boldsymbol{F}^j\right),$$

where everything is known in advance and therefore it can be computed and translated to the right-hand side of the resulting algebraic equation.

4.2.2 Matrix Form of the Term $T4$

Proceeding in the same way,

$$T4 = \int_{\bar{\Gamma}} \int_{\Omega} \nabla_s (\boldsymbol{R}^*(\boldsymbol{x}) \cdot S(\boldsymbol{s})) : \mathbf{C} : \nabla^T \left(\sum_{I=1}^{\mathtt{nprev}} \boldsymbol{F}^I(\boldsymbol{x}) \cdot \boldsymbol{G}^I(\boldsymbol{s}) \right) \cdot \nabla \left(\sum_{j=1}^{n} \boldsymbol{F}^j(\boldsymbol{x}) \cdot G^j(\boldsymbol{s}) + \boldsymbol{R}(\boldsymbol{x}) \cdot S(\boldsymbol{s}) \right) d\Omega d\bar{\Gamma}.$$

Note that the nodal values for functions $\boldsymbol{F}^I(\boldsymbol{x})$ and $\boldsymbol{G}^I(\boldsymbol{s})$ are named in the code reproduced below as `Fprev` and `Gprev`, respectively.

Again, part of the terms of $T4$ involve unknown functions, while the others involve already known terms, and will therefore translated to the force vector in the resulting algebraic equation:

$$\begin{aligned} T4^K &= \int_{\bar{\Gamma}} \int_{\Omega} \nabla_s (\boldsymbol{R}^*(\boldsymbol{x}) \cdot S(\boldsymbol{s})) : \mathbf{C} : \nabla^T \left(\sum_{I=1}^{\mathtt{nprev}} \boldsymbol{F}^I(\boldsymbol{x}) \cdot G^I(\boldsymbol{s}) \right) \cdot \nabla (\boldsymbol{R}(\boldsymbol{x}) \cdot S(\boldsymbol{s})) d\Omega d\bar{\Gamma} \\ &= \sum_{I=1}^{\mathtt{nprev}} \int_{\bar{\Gamma}} S(\boldsymbol{s}) \cdot G^I(\boldsymbol{s}) \cdot S(\boldsymbol{s}) d\bar{\Gamma} \int_{\Omega} \nabla_s (\boldsymbol{R}^*(\boldsymbol{x})) : \mathbf{C} : \nabla^T (\boldsymbol{F}^I(\boldsymbol{x})) \cdot \nabla (\boldsymbol{R}(\boldsymbol{x})) d\Omega, \end{aligned}$$

and

$$\begin{aligned} T4^F &= \int_{\bar{\Gamma}} \int_{\Omega} \nabla_s (\boldsymbol{R}^*(\boldsymbol{x}) \cdot S(\boldsymbol{s})) : \mathbf{C} : \nabla^T \left(\sum_{I=1}^{\mathtt{nprev}} \boldsymbol{F}^I(\boldsymbol{x}) \cdot G^I(\boldsymbol{s}) \right) \cdot \nabla \left(\sum_{j=1}^{n} \boldsymbol{F}^j(\boldsymbol{x} \right) \cdot G^j(\boldsymbol{s})) d\Omega d\bar{\Gamma} \\ &= \sum_{I=1}^{\mathtt{nprev}} \sum_{j=1}^{n} \int_{\bar{\Gamma}} S(\boldsymbol{s}) \cdot G^I(\boldsymbol{s}) \cdot G^j(\boldsymbol{s}) d\bar{\Gamma} \int_{\Omega} \nabla_s (\boldsymbol{R}^*(\boldsymbol{x})) : \mathbf{C} : \nabla^T (\boldsymbol{F}^I(\boldsymbol{x})) \cdot \nabla (\boldsymbol{F}^j(\boldsymbol{x})) d\Omega. \end{aligned}$$

By using gain linear finite element shape functions and the same notation as in the previous paragraph, we arrive at

$$\begin{aligned} T4^K &= \sum_{I=1}^{\mathtt{nprev}} \left(\boldsymbol{s}^T \left[\int_{\bar{\Gamma}} \boldsymbol{M}^T \boldsymbol{M} \boldsymbol{G}^I \boldsymbol{M} d\bar{\Gamma} \right] \boldsymbol{s} \right) \left(\boldsymbol{R}^{*T} \left[\int_{\Omega} \boldsymbol{B}^T \mathbf{C} \widetilde{\boldsymbol{A}}_{\boldsymbol{F}^I(\boldsymbol{x})} \widetilde{\boldsymbol{G}} d\Omega \right] \boldsymbol{R} \right) \\ &= \sum_{I=1}^{\mathtt{nprev}} \left(\boldsymbol{s}^T \boldsymbol{s}_{2I} \boldsymbol{s} \right) \left(\boldsymbol{R}^{*T} \boldsymbol{R}_{2I} \boldsymbol{R} \right), \end{aligned}$$

while

$$T4^F = \sum_{I=1}^{\mathtt{nprev}} \sum_{j=1}^{n} \left(s^T s_{2I} G^j\right) \left(R^{*T} R_{2I} F^j\right),$$

where

$$\widetilde{A}_F = \widetilde{A}(F) = \begin{bmatrix} f_{x,x} & f_{y,x} & f_{z,x} & 0 & 0 & 0 & 0 & 0 & 0 \\ 0 & 0 & 0 & f_{x,y} & f_{y,y} & f_{z,y} & 0 & 0 & 0 \\ 0 & 0 & 0 & 0 & 0 & 0 & f_{x,z} & f_{y,z} & f_{z,z} \\ f_{x,y} & f_{y,y} & f_{z,y} & f_{x,x} & f_{y,x} & f_{z,x} & 0 & 0 & 0 \\ 0 & 0 & 0 & f_{x,z} & f_{y,z} & f_{z,z} & f_{x,y} & f_{y,y} & f_{z,y} \\ f_{x,z} & f_{y,z} & f_{z,z} & 0 & 0 & 0 & f_{x,x} & f_{y,x} & f_{z,x} \end{bmatrix},$$

and

$$\widetilde{G} = \begin{bmatrix} \varphi_{1,x}(x) & 0 & 0 & \varphi_{2,x}(x) & 0 & 0 & \cdots & 0 \\ 0 & \varphi_{1,x}(x) & 0 & 0 & \varphi_{2,x}(x) & 0 & \cdots & 0 \\ 0 & 0 & \varphi_{1,x}(x) & 0 & 0 & \varphi_{2,x}(x) & \cdots & \varphi_{N,x}(x) \\ \varphi_{1,y}(x) & 0 & 0 & \varphi_{2,y}(x) & 0 & 0 & \cdots & 0 \\ 0 & \varphi_{1,y}(x) & 0 & 0 & \varphi_{2,y}(x) & 0 & \cdots & 0 \\ 0 & 0 & \varphi_{1,y}(x) & 0 & 0 & \varphi_{2,y}(x) & \cdots & \varphi_{N,y}(x) \\ \varphi_{1,z}(x) & 0 & 0 & \varphi_{2,z}(x) & 0 & 0 & \cdots & 0 \\ 0 & \varphi_{1,z}(x) & 0 & 0 & \varphi_{2,z}(x) & 0 & \cdots & 0 \\ 0 & 0 & \varphi_{1,z}(x) & 0 & 0 & \varphi_{2,z}(x) & \cdots & \varphi_{N,z}(x) \end{bmatrix}.$$

4.2.3 Matrix Form of the Term $T6$

The term $T6$ is indedd very similar to the $T4$ one, but changing the position of the gradient operators.

$$T6 = \int_{\bar{\Gamma}} \int_{\Omega} \nabla^T (R^*(x) \cdot S(s)) \cdot \nabla \left(\sum_{I=1}^{\mathtt{nprev}} F^I(x \right) \cdot G^I(s)) : \mathbf{C} : \nabla_s$$
$$\times \left(\sum_{j=1}^{n} F^j(x) \cdot G^j(s) + R(x) \cdot S(s) \right) d\Omega d\bar{\Gamma}.$$

By decomposing it, as in $T4$, in known and unknown parts, we obtain

$$T6 = T6^K + T6^F = \sum_{I=1}^{\mathtt{nprev}} \left(s^T s_{2I} s\right) \left(R^{*T} R_{2I}^T R\right) + \sum_{I=1}^{\mathtt{nprev}} \sum_{j=1}^{n} \left(s^T s_{2I} G^j\right) \left(R^{*T} R_{2I}^T F^j\right).$$

Again, the term $T6^K$ will contribute to the stiffness matrix, while $T6^F$ will contribute to the right-hand side of the equation, usually referred to as "force vector".

4.2.4 Matrix Form for the Term $T8$

In this case, the term $T8$ will look like

$$T8 = \int_{\bar{\Gamma}} \int_{\Omega} \nabla^T (\boldsymbol{R}^*(\boldsymbol{x}) \cdot S(s)) \cdot \nabla \left(\sum_{I=1}^{\texttt{nprev}} \boldsymbol{F}^I(\boldsymbol{x}) \cdot G^I(s) \right) : \mathbf{C} : \nabla^T \\ \times \left(\sum_{J=1}^{\texttt{nprev}} \boldsymbol{F}^J(\boldsymbol{x}) \cdot G^J(s) \right) \cdot \nabla \left(\sum_{j=1}^{n} \boldsymbol{F}^j(\boldsymbol{x}) \cdot G^j(s) + \boldsymbol{R}(\boldsymbol{x}) \cdot S(s) \right) d\Omega d\bar{\Gamma}.$$

Again, the term is decomposed into parts containing the unknown and known parts, which will contribute, respectively, to the stiffness matrix and the force vector:

$$T8^K = \int_{\bar{\Gamma}} \int_{\Omega} \nabla^T (\boldsymbol{R}^*(\boldsymbol{x}) \cdot S(s)) \cdot \nabla \left(\sum_{I=1}^{\texttt{nprev}} \boldsymbol{F}^I(\boldsymbol{x}) \cdot G^I(s) \right) : \\ \times \mathbf{C} : \nabla^T \left(\sum_{J=1}^{\texttt{nprev}} \boldsymbol{F}^J(\boldsymbol{x}) \cdot G^J(s) \right) \cdot \nabla (\boldsymbol{R}(\boldsymbol{x}) \cdot S(s)) d\Omega d\bar{\Gamma} \\ = \sum_{I=1}^{\texttt{nprev}} \sum_{J=1}^{\texttt{nprev}} \int_{\bar{\Gamma}} S(s) G^I(s) G^J(s) S(s) d\bar{\Gamma} \int_{\Omega} \nabla^T \boldsymbol{R}^*(\boldsymbol{x}) \cdot \nabla \boldsymbol{F}^I(\boldsymbol{x}) : \\ \times \mathbf{C} : \nabla^T \boldsymbol{F}^J(\boldsymbol{x}) \cdot \nabla \boldsymbol{R}(\boldsymbol{x}) d\Omega.$$

Conversely, the term $T8^F$ reads

$$T8^F = \int_{\bar{\Gamma}} \int_{\Omega} \nabla^T (\boldsymbol{R}^*(\boldsymbol{x}) \cdot S(s)) \cdot \nabla \left(\sum_{I=1}^{\texttt{nprev}} \boldsymbol{F}^I(\boldsymbol{x}) \cdot G^I(s) \right) : \\ \times \mathbf{C} : \nabla^T \left(\sum_{J=1}^{\texttt{nprev}} \boldsymbol{F}^J(\boldsymbol{x}) \cdot G^J(s) \right) \cdot \nabla \left(\sum_{j=1}^{n} \boldsymbol{F}^j(\boldsymbol{x} \right) \cdot G^j(s)) d\Omega d\bar{\Gamma} \\ = \sum_{I=1}^{\texttt{nprev}} \sum_{J=1}^{\texttt{nprev}} \sum_{j=1}^{n} \int_{\bar{\Gamma}} S(s) G^I(s) G^J(s) G^j(s) d\bar{\Gamma} \int_{\Omega} \nabla^T \boldsymbol{R}^*(\boldsymbol{x}) \cdot \nabla \boldsymbol{F}^I(\boldsymbol{x}) : \\ \times \mathbf{C} : \nabla^T \boldsymbol{F}^J(\boldsymbol{x}) \cdot \nabla \boldsymbol{F}^j(\boldsymbol{x}) d\Omega.$$

Finally, both terms are approximated with the help of linear finite element shape functions, giving rise to

$$T8^K = \sum_{I=1}^{\texttt{nprev}} \sum_{J=1}^{\texttt{nprev}} \left(\boldsymbol{s}^T \left[\int_{\bar{\Gamma}} \boldsymbol{M}^T \boldsymbol{M} \boldsymbol{G}^I \boldsymbol{M} \boldsymbol{G}^J \boldsymbol{M} d\bar{\Gamma} \right] \boldsymbol{s} \right) \times \left(\boldsymbol{R}^{*T} \left[\int_{\Omega} \widetilde{\boldsymbol{G}}^T \widetilde{\boldsymbol{A}}^T_{\boldsymbol{F}^I(x)} \mathbf{C} \widetilde{\boldsymbol{A}}_{\boldsymbol{F}^J(x)} \widetilde{\boldsymbol{G}} d\Omega \right] \boldsymbol{R} \right),$$

or, equivalently,

$$T8^K = \sum_{I=1}^{\texttt{nprev}} \sum_{J=1}^{\texttt{nprev}} \left(\boldsymbol{s}^T \boldsymbol{s}_{3IJ} \boldsymbol{s} \right) \left(\boldsymbol{R}^{*T} \boldsymbol{R}_{4IJ} \boldsymbol{R} \right),$$

while the contribution to the force vector will now be

$$T8^F = \sum_{I=1}^{\texttt{nprev}} \sum_{J=1}^{\texttt{nprev}} \sum_{j=1}^{n} \left(\boldsymbol{s}^T \boldsymbol{s}_{3IJ} \boldsymbol{G}^j \right) \left(\boldsymbol{R}^{*T} \boldsymbol{R}_{4IJ} \boldsymbol{F}^j \right).$$

4.2.5 Matrix Form of the Term $T9$

For the term $T9$ we follow similar guidelines as in previous cases,

$$T9 = \int_{\bar{\Gamma}} \int_{\Omega} \nabla^T (\boldsymbol{R}^*(\boldsymbol{x}) \cdot S(\boldsymbol{s})) \cdot \nabla \left(\sum_{j=1}^{n} \boldsymbol{F}^j(\boldsymbol{x}) \cdot G^j(\boldsymbol{s}) + \boldsymbol{R}(\boldsymbol{x}) \cdot S(\boldsymbol{s}) \right) : \times \mathbf{C} : \nabla_s \left(\sum_{I=1}^{\texttt{nprev}} \boldsymbol{F}^I(\boldsymbol{x}) \cdot G^I(\boldsymbol{s}) \right) d\Omega d\bar{\Gamma}.$$

Again, the term is decomposed into contributions to the stiffness matrix and the force vector:

$$\begin{aligned} T9^K &= \int_{\bar{\Gamma}} \int_{\Omega} \nabla^T (\boldsymbol{R}^*(\boldsymbol{x}) \cdot S(\boldsymbol{s})) \cdot \nabla (\boldsymbol{R}(\boldsymbol{x}) \cdot S(\boldsymbol{s})) : \mathbf{C} : \nabla_s \left(\sum_{I=1}^{\texttt{nprev}} \boldsymbol{F}^I(\boldsymbol{x}) \cdot G^I(\boldsymbol{s}) \right) d\Omega d\bar{\Gamma} \\ &= \sum_{I=1}^{\texttt{nprev}} \int_{\bar{\Gamma}} S(s) S(s) G^I(s) d\bar{\Gamma} \int_{\Omega} \nabla^T \boldsymbol{R}^*(\boldsymbol{x}) \cdot \nabla \boldsymbol{R}(\boldsymbol{x}) : \mathbf{C} : \nabla_s \boldsymbol{F}^I(\boldsymbol{x}) d\Omega, \end{aligned}$$

and

$$T9^F = \int_{\bar{\Gamma}} \int_{\Omega} \nabla^T (\boldsymbol{R}^*(\boldsymbol{x}) \cdot S(\boldsymbol{s})) \cdot \nabla \left(\sum_{j=1}^{n} \boldsymbol{F}^j(\boldsymbol{x}) \cdot G^j(\boldsymbol{s}) \right) :$$
$$\times \mathbf{C} : \nabla_s (\sum_{l=1}^{\texttt{nprev}} \boldsymbol{F}^l(\boldsymbol{x}) \cdot G^l(\boldsymbol{s})) d\Omega d\bar{\Gamma}$$
$$= \sum_{l=1}^{\texttt{nprev}} \sum_{j=1}^{n} \int_{\bar{\Gamma}} S(\boldsymbol{s}) G^j(\boldsymbol{s}) G^l(\boldsymbol{s}) d\bar{\Gamma} \int_{\Omega} \nabla^T \boldsymbol{R}^*(\boldsymbol{x}) \cdot \nabla \boldsymbol{F}^j(\boldsymbol{x}) :$$
$$\times \mathbf{C} : \nabla_s \boldsymbol{F}^l(\boldsymbol{x}) d\Omega.$$

Finally, we need to approximate all the involved functions by means of (in this case, linear) finite element shape functions, leading to

$$T9^K = \sum_{l=1}^{\texttt{nprev}} \left(\boldsymbol{s}^T \left[\int_{\bar{\Gamma}} \boldsymbol{M}^T \boldsymbol{M} \boldsymbol{G}^l \boldsymbol{M} d\bar{\Gamma} \right] \boldsymbol{s} \right) \left(\boldsymbol{R}^{*T} \left[\int_{\Omega} \tilde{\boldsymbol{G}}^T f_{1(\mathbf{C}, \nabla_s \boldsymbol{F}^l)} \tilde{\boldsymbol{G}} d\Omega \right] \boldsymbol{R} \right)$$
$$= \sum_{l=1}^{\texttt{nprev}} \left(\boldsymbol{s}^T \boldsymbol{s}_{2l} \boldsymbol{s} \right) \left(\boldsymbol{R}^{*T} \boldsymbol{R}_{3l} \boldsymbol{R} \right), \quad (4.8)$$

and

$$T9^F = \sum_{l=1}^{\texttt{nprev}} \sum_{j=1}^{n} \left(\boldsymbol{s}^T \boldsymbol{s}_{2l} \boldsymbol{G}^j \right) \left(\boldsymbol{R}^{*T} \boldsymbol{R}_{3l} \boldsymbol{F}^j \right).$$

The term $f_{1(\mathbf{C}, \nabla_s \boldsymbol{F}^l)}$ in Eq. (4.8) is in fact a matrix containing the components of the vector obtained after multiplication of $\mathbf{C}$ times $\nabla_s \boldsymbol{F}^l$. Indedd, if

$$\boldsymbol{x} = [X_1, X_2, \ldots X_6] = \mathbf{C} \cdot \nabla_s \boldsymbol{F}^l(\boldsymbol{x}) = \mathbf{C} \boldsymbol{B} \boldsymbol{F}^l,$$

then

$$f_{1(\mathbf{C}, \nabla_s \boldsymbol{F}^l)} = \begin{bmatrix} X_1 & 0 & 0 & X_4 & 0 & 0 & X_5 & 0 & 0 \\ 0 & X_1 & 0 & 0 & X_4 & 0 & 0 & X_5 & 0 \\ 0 & 0 & X_1 & 0 & 0 & X_4 & 0 & 0 & X_5 \\ X_4 & 0 & 0 & X_2 & 0 & 0 & X_6 & 0 & 0 \\ 0 & X_4 & 0 & 0 & X_2 & 0 & 0 & X_6 & 0 \\ 0 & 0 & X_4 & 0 & 0 & X_2 & 0 & 0 & X_6 \\ X_5 & 0 & 0 & X_6 & 0 & 0 & X_3 & 0 & 0 \\ 0 & X_5 & 0 & 0 & X_6 & 0 & 0 & X_3 & 0 \\ 0 & 0 & X_5 & 0 & 0 & X_6 & 0 & 0 & X_3 \end{bmatrix}.$$

4.2.6 Matrix Form of the Term $T10$

As can be noticed from Eq. (4.6), the term $T10$ has an expression

$$T10 = \int_{\bar{\Gamma}}\int_{\Omega} \nabla^T (\boldsymbol{R}^*(\boldsymbol{x}) \cdot S(\boldsymbol{s})) \cdot \nabla \left(\sum_{j=1}^{n} \boldsymbol{F}^j(\boldsymbol{x}) \cdot G^j(\boldsymbol{s}) + \boldsymbol{R}(\boldsymbol{x}) \cdot S(\boldsymbol{s}) \right) : \mathbf{C}$$
$$\times : \frac{1}{2} \nabla^T \left(\sum_{I=1}^{\text{nprev}} \boldsymbol{F}^I(\boldsymbol{x}) \cdot G^I(\boldsymbol{s}) \right) \nabla \left(\sum_{I=1}^{\text{nprev}} \boldsymbol{F}^I(\boldsymbol{x}) \cdot G^I(\boldsymbol{s}) \right) d\Omega d\bar{\Gamma}.$$

By decomposing it into stiffness and force vector contributions,

$$\begin{aligned} T10^K &= \int_{\bar{\Gamma}}\int_{\Omega} \nabla^T (\boldsymbol{R}^*(\boldsymbol{x}) \cdot S(\boldsymbol{s})) \cdot \nabla (\boldsymbol{R}(\boldsymbol{x}) \cdot S(\boldsymbol{s})) : \\ &\quad \times \mathbf{C} : \frac{1}{2} \nabla^T \left(\sum_{I=1}^{\text{nprev}} \boldsymbol{F}^I(\boldsymbol{x}) \cdot G^I(\boldsymbol{s}) \right) \nabla \left(\sum_{J=1}^{\text{nprev}} \boldsymbol{F}^J(\boldsymbol{x}) \cdot G^J(\boldsymbol{s}) \right) d\Omega d\bar{\Gamma} \\ &= \frac{1}{2} \sum_{I=1}^{\text{nprev}} \sum_{J=1}^{\text{nprev}} \int_{\bar{\Gamma}} S(\boldsymbol{s}) S(\boldsymbol{s}) G^I(\boldsymbol{s}) G^J(\boldsymbol{s}) d\bar{\Gamma} \int_{\Omega} \nabla^T \boldsymbol{R}^*(\boldsymbol{x}) \cdot \nabla \boldsymbol{R}(\boldsymbol{x}) : \\ &\quad \times \mathbf{C} : \nabla^T \boldsymbol{F}^I(\boldsymbol{x}) \cdot \nabla \boldsymbol{F}^J(\boldsymbol{x}) d\Omega, \end{aligned}$$

with

$$\begin{aligned} T10^F &= \int_{\bar{\Gamma}}\int_{\Omega} \nabla^T (\boldsymbol{R}^*(\boldsymbol{x}) \cdot S(\boldsymbol{s})) \cdot \nabla \left(\sum_{j=1}^{n} \boldsymbol{F}^j(\boldsymbol{x}) \cdot G^j(\boldsymbol{s}) \right) : \\ &\quad \times \mathbf{C} : \frac{1}{2} \nabla^T \left(\sum_{I=1}^{\text{nprev}} \boldsymbol{F}^I(\boldsymbol{x}) \cdot G^I(\boldsymbol{s}) \right) \nabla \left(\sum_{J=1}^{\text{nprev}} \boldsymbol{F}^J(\boldsymbol{x}) \cdot G^J(\boldsymbol{s}) \right) d\Omega d\bar{\Gamma} \\ &= \frac{1}{2} \sum_{I=1}^{\text{nprev}} \sum_{J=1}^{\text{nprev}} \sum_{j=1}^{n} \int_{\bar{\Gamma}} S(\boldsymbol{s}) G^j(\boldsymbol{s}) G^I(\boldsymbol{s}) G^J(\boldsymbol{s}) d\bar{\Gamma} \int_{\Omega} \nabla^T \boldsymbol{R}^*(\boldsymbol{x}) \cdot \nabla \boldsymbol{F}^j(\boldsymbol{x}) : \\ &\quad \times \mathbf{C} : \nabla^T \boldsymbol{F}^I(\boldsymbol{x}) \cdot \nabla \boldsymbol{F}^J(\boldsymbol{x}) d\Omega. \end{aligned}$$

Finally, after finite element approximation, we obtain, in pseudo-vectorial notation,

$$\begin{aligned} T10^K &= \frac{1}{2} \sum_{I=1}^{\text{nprev}} \sum_{J=1}^{\text{nprev}} \left(\boldsymbol{s}^T \left[\int_{\bar{\Gamma}} \boldsymbol{M}^T \boldsymbol{M} \boldsymbol{G}^I \boldsymbol{M} \boldsymbol{G}^J \boldsymbol{M} d\bar{\Gamma} \right] \boldsymbol{s} \right) \\ &\quad \times \left(\boldsymbol{R}^{*T} \left[\int_{\Omega} \widetilde{\boldsymbol{G}}^T f_{1(\mathbf{C}, \nabla^T \boldsymbol{F}^I \nabla \boldsymbol{F}^J)} \widetilde{\boldsymbol{G}} d\Omega \right] \boldsymbol{R} \right), \end{aligned}$$

or, equivalently,

$$T10^K = \frac{1}{2} \sum_{I=1}^{\mathtt{nprev}} \sum_{J=1}^{\mathtt{nprev}} \left(\boldsymbol{S}^T \boldsymbol{s}_{31J} \boldsymbol{S} \right) \left(\boldsymbol{R}^{*T} \boldsymbol{r}_{51J} \boldsymbol{R} \right).$$

The term contributing to the force vector will finally read

$$T10^F = \frac{1}{2} \sum_{I=1}^{\mathtt{nprev}} \sum_{J=1}^{\mathtt{nprev}} \sum_{j=1}^{n} \left(\boldsymbol{S}^T \boldsymbol{s}_{31J} \boldsymbol{G}^j \right) \left(\boldsymbol{R}^{*T} \boldsymbol{r}_{51J} \boldsymbol{F}^j \right).$$

4.2.7 *Final Comments*

We have detailed one of the two iterations of the fixed point algorithm. The case in which we look for the function S is essentially identical to what has been here described.

As can be noticed, the method here developed is purely explicit, in the sense that matrices $\boldsymbol{r}$ and $\boldsymbol{s}$ contain only terms for which a closed-form expression is known. There is no need to iterate to determine their precise expression. There are, indeed, iterations in the fixed point, alternating directions algorithm. These are due to the seek for a product of functions, which renders actually a non-linear problem.

4.3 Matlab Code

As in previous chapters, the code begins by the `main.m` routine. The routine reads the model data (nodal coordinates and connectivity) from the file `gcoordBeam.dat` and `conecBeam.dat` files, respectively. Open these files with a standard plain text editor to see their structure.

The problem to solve represents a cantilever beam subjected to bending loads. By typing

```
» trisurf(tri,coors(:,1),coors(:,2),coors(:,3));
» axis equal
```

a plot of the beam model, see Fig. 4.1, is depicted. It is essentially the same employed in Chap. 5.

The problem has been solved by means of a very crude tetrahedral mesh of 3 × 3 × 16 nodes. Following the structure of the algorithm just introduced, see Eq. (4.5), the load is applied in a sequence of 5 increments. These increments produce a series of 5 terms for the first increment and 10 for the subsequent ones.

The instruction `save('WorkSpacePGD_Hyperelastic.mat');` saves all the results (in essence, the F^j and G^j functions in Eq. (4.7)) and ends what we call the *off-line* phase of the simulation. These results need not to be calculated

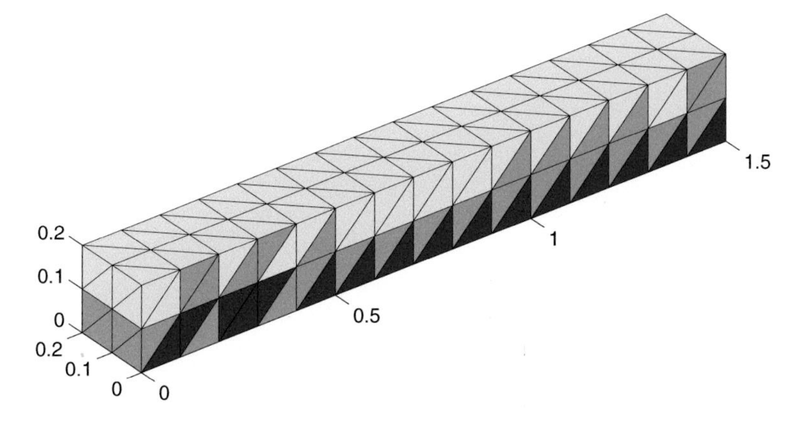

Fig. 4.1 Finite element mesh for the beam bending problem. The beam is assumed to be encastred at $y = 0$

again and therefore must be stored in memory. All subsequent simulations need only to read them in what we call the *on-line* phase of the simulation.

By the `CoordSelect = [1.5 0 0];` instruction we begin the on-line phase by indicating the program the particular position in which the load is applied. Feel free to change it to any point in the upper surface of the beam. In Chap. 3 we saw how to do this interactively, by selecting the loading point directly on the screen with the help of the mouse.

```
%
%                        PGD Code for hyperelasticity
%                        I. Alfaro, D. Gonzalez, E. Cueto
%                          Universidad de Zaragoza
%                            AMB-I3A Dec 2015
%
clear all; clc; format long g; close all;
%
% VARIABLES
%
% Global variables.
global coords tet tri tf
global Fprev Gprev num_iter_prev
global E nu
global s1 vx vp
E = 2.1e11; nu = 0.25; % Material (Young Modulus and Poisson Coef).
Modulus_init = 30e6;% Total force Modulus.
nincr = 5; % Number of load increments.
Modulus = Modulus_init/nincr; % Force on each load increment.
TOL = 0.05; % Tolerance.
iter = zeros(1); % # of iterations needed on enrichment.m function
num_max_iter = 10; % Max. # of summands for the approach on each load incr.
%
% GEOMETRY AND BOUNDARY CONDITIONS
%
% Nodes and elements readed from external files.
coords = load('gcoordBeam.dat'); % Nodal coordinates.
tet = load('conecBeam.dat'); % Connectivity list.
Ind = 1:size(coords,1); % List of nodes.
bcnode = Ind(coords(:,1)==min(coords(:,1))); % Boundary: Fix left side.
IndBcnode = sort([3*(bcnode-1)+1 3*(bcnode-1)+2 3*bcnode]); % D.o.f. BCs.
% Load can be applied in every node of the surface
% Make  use of triangulation MatLab function to obtain boundary surface.
```

```
TR = triangulation(tet, coords);
% Connectivity of tf corresponds to the node number of the whole domain
% Connectivity of tri corresponds to the node number of the free boundary
[tf] = freeBoundary(TR); % Dependent of 3D geometry of the boundary.
[tri,coors] = freeBoundary(TR); % Independent triangulation of boundary.
IndS = 1:size(coors,1); ncoors = numel(IndS);
%
% ALLOCATION OF MATRICES AND VECTORS
%
% Vector solutions of all previous increments (cumulative).
Fprev = zeros(numel(coords),1); % Vector solution for space.
Gprev = zeros(size(coors,1),1); % Vector solution for load.
num_iter_prev = 0; % total # of summands.

% Loop on every load increment
for incr=1:nincr
    fprintf(1,'load step %d\n',incr);
    %
    % INITIALIZATION OF MATRICES AND VECTORS
    %
    F = zeros(numel(coords),1);% For space in current increment.
    R = zeros(numel(coords),1);% For load in current increment.
    G = zeros(size(coors,1),1),
    S = zeros(size(coors,1),1);
    num_iter = 0;
    Error_iter = 1.0;
    Aprt = 0;
    %
    % STIFNESS AND MASS MATRICES COMPUTATION
    %
    fem3DHyperelastic;
    coorp = 1:size(coors,1); % We consider each position like cases of load
    elemstiffHyperelastic(coorp);
    %
    % SOURCE
    %
    % Identifying local nodes of the loaded surface on the global connectivity.
    % To obtain that: coords(IndL,:)-coors = zeros(nn2,1).
    [trash,trash2,xj] = intersect(IndS,tri(:)); % TRI Local connectivity.
    IndL = tf(xj); % TF Global connectivity of the loaded surface.
    DOFLoaded = 3*IndL; % Consider vertical load on the top of the beam.
    vx = zeros(numel(coords),ncoors);
    vx(DOFLoaded,:) = -Modulus.*eye(ncoors); % Space terms for the source.
    vp = eye(ncoors); vp = s1*vp; % Load terms for the source.
    %
    % ENRICHMENT OF THE APPROXIMATION, LOOKING FOR R AND S
    %
    while Error_iter > TOL && num_iter < num_max_iter
        num_iter = num_iter + 1;
        S0 = rand(size(coors,1),1); % Initial guess for S.
        %
        % ENRICHMENT STEP
        %
        [R,S,iter(num_iter)] = enrichment(S0,F,G,TOL,IndBcnode,num_iter);
        F(:,num_iter) = R;
        G(:,num_iter) = S;
        %
        % STOPPING CRITERION
        %
        Error_iter = norm(F(:,num_iter)*G(:,num_iter)');
        Aprt = max(Aprt,sqrt(Error_iter));
        Error_iter = sqrt(Error_iter)/Aprt;
        fprintf(1,'%dst summand in %d iterations with a weight of %f\n',...
            num_iter,iter(num_iter),Error_iter);
    end
    if num_iter>1, num_iter = num_iter - 1; end % last sum is negligible.
    % Before going to the next increment, add current vector solutions to
    % previous vector solutions.
    Fprev(:,num_iter_prev+1:num_iter_prev+num_iter) = F(:,1:num_iter);
    Gprev(:,num_iter_prev+1:num_iter_prev+num_iter) = G(:,1:num_iter);
    num_iter_prev = num_iter_prev + num_iter;
end
```

```
F = Fprev; G = Gprev; num_iter = num_iter_prev;
fprintf(1,'PGD␣off-line␣Process␣exited␣normally\n\n');
save('WorkSpacePGD_Hyperelastic.mat');
%
% POST-PROCESSING
%
CoordSelect = [1.5 0 0]; % coordenates where load is applied.
Cx = CoordSelect(1); Cy = CoordSelect(2); Cz = CoordSelect(3);
[trash,LoadPos] = min(dist(coors,[Cx;Cy;Cz])); % closest node to load.
disp = zeros(numel(coords),1);
for i1=1:num_iter
    disp = disp + F(:,i1).*G(LoadPos,i1);
end
cdx = coords(:,1) + disp(1:3:end);
cdy = coords(:,2) + disp(2:3:end);
cdz = coords(:,3) + disp(3:3:end);
figure(1);
trepmod = TriRep(tet, [cdx cdy cdz]); % deformed domain.
[trimod, Xbmod] = freeBoundary(trepmod);
nc = 1:size(coors,1);
[trash,trash2,xj] = intersect(nc,tri(:));
Ind = tf(xj);
trisurf(trimod, Xbmod(:,1),Xbmod(:,2),Xbmod(:,3),disp(3*Ind),...
    'FaceAlpha',0.8); % plot the deformed surface.
axis equal
fprintf(1,'max␣displ␣%f\n',min(disp)); %  print maximum displacement.
fprintf(1,'\n\n##########␣End␣of␣simulation ##########\n\n');
```

The `main.m` routine first calls to the `fem3DHyperelastic` subroutine, which corresponds roughly to a standard three-dimensional finite element program. In this case, we employ simple linear tetrahedrons and Hammer quadrature rules. Different matrices of the terms $T2$, $T4$, ..., are also calculated here.

```
function fem3DHyperelastic()
% function femSpace3DHyperelastic
% Computes r1, r2I, r3I, r4IJ and r5IJ matrices
% Universidad de Zaragoza - 2015
global E nu
global coords tet num_iter_prev Fprev
global r1 r2 r3 r5 r4

dof = 3; % Degrees of freedom per node
numNodes = size(coords,1);
numTet = size(tet,1);

% r1 is a sparse matrix which is calculated only at the first time
% increment.
% r2 is a cell containing all matrices r2I. r2{I}=r2I (same for r3)
% r4 is a cell containing all matrices r4IJ (same for r5)
% Nic is the number of increments in wich matrices are already calculated.
% Only new matrices are calculated.
NIc = 0;
if num_iter_prev > 0
    NIc = size(r2,1); % Number of matrices r2 already calculated.
    if isempty(r2)==1, NIc = 0; end
end
if NIc > 0
    % Save previously calculated matrices on auxiliary variables.
    r2_aux = r2; r3_aux = r3; r4_aux = r4; r5_aux = r5;
end
%
% ALLOCATE MEMORY AND INICIALIZE
%
if num_iter_prev == 0
    r1 = sparse(dof*numNodes,dof*numNodes);
end
r2 = cell(num_iter_prev,1);
r4 = cell(num_iter_prev*num_iter_prev,1);
for i1=1:num_iter_prev, r2{i1} = sparse(dof*numNodes,dof*numNodes); end
for i1=1:num_iter_prev*num_iter_prev
```

```
    r4{i1} = sparse(dof*numNodes,dof*numNodes);
end
r3 = r2; r5 = r4;
%
% ADD ALREADY CALCULATED MATRICES
%
if NIc > 0
    r2(1:NIc) = r2_aux; r3(1:NIc) = r3_aux;
    NewIncre = num_iter_prev-NIc;
    if NewIncre > 0
        IndMat = zeros(NIc*NIc,NIc); IndMat(:,1) = [1:NIc*NIc]';
        for i1=2:NIc
            IndMat(:,i1) = IndMat(:,i1-1) + NewIncre.*ones(size(IndMat,1),1);
        end
        IndMat = reshape(IndMat(:),[NIc,NIc*NIc]);
        IM = IndMat(:,1:NIc+1:end);
        r4(IM(:)) = r4_aux; r5(IM(:)) = r5_aux;
    else
        r4 = r4_aux; r5 = r5_aux;
    end
end
%
% CALCULATE NEW MATRICES
%
%
% MATERIAL MATRIX
%
D = zeros(6); cte = E*(1-nu)/(1+nu)/(1-2*nu);
D(1) = cte; D(8) = D(1); D(15) = D(1);
D(2) = cte*nu/(1-nu); D(3) = D(2); D(7) = D(2); D(9) = D(2); D(13) = D(2);
D(14) = D(2); D(22) = cte*(1-2*nu)/2/(1-nu); D(29) = D(22); D(36) = D(22);
%
% INTEGRATION POINTS: 4 HAMMER POINTS
%
sg = zeros(3,4); wg = 1./24.*ones(4,1); nph = numel(wg);
a = (5.0 - sqrt(5))/20.0; b= (5.0 + 3.0*sqrt(5))/20.0;
sg(:,1) = [a; a; a]; sg(:,2) = [a; a; b];
sg(:,3) = [a; b; a];  sg(:,4) = [b; a; a];
%
% ELEMENT LOOP
%
for i1=1:numTet
    elnodes = tet(i1,:);
    xcoord = coords(elnodes,:);
    % System degrees of freedom associated with each element.
    index = [3*elnodes-2;3*elnodes-1;3*elnodes];
    index = reshape(index,1,4*dof);
    %
    % JACOBIAN
    %
    v1 = xcoord(1,:)-xcoord(2,:); v2 = xcoord(2,:)-xcoord(3,:);
    v3 = xcoord(3,:)-xcoord(4,:);
    jcob = abs(det([v1;v2;v3]));
    %
    % SHAPE FUNCTION CONSTANTS
    %
    a1 = det([xcoord(2,:); xcoord(3,:); xcoord(4,:)]);
    a2 = -det([xcoord(1,:); xcoord(3,:); xcoord(4,:)]);
    a3 = det([xcoord(1,:); xcoord(2,:); xcoord(4,:)]);
    a4 = -det([xcoord(1,:); xcoord(2,:); xcoord(3,:)]);
    b1 = -det([1 xcoord(2,2:end); 1 xcoord(3,2:end); 1 xcoord(4,2:end)]);
    b2 = det([1 xcoord(1,2:end); 1 xcoord(3,2:end); 1 xcoord(4,2:end)]);
    b3 = -det([1 xcoord(1,2:end); 1 xcoord(2,2:end); 1 xcoord(4,2:end)]);
    b4 = det([1 xcoord(1,2:end); 1 xcoord(2,2:end); 1 xcoord(3,2:end)]);
    c1 = det([1 xcoord(2,1) xcoord(2,end); 1 xcoord(3,1) xcoord(3,end);...
        1 xcoord(4,1) xcoord(4,end)]);
    c2 = -det([1 xcoord(1,1) xcoord(1,end); 1 xcoord(3,1) xcoord(3,end);...
        1 xcoord(4,1) xcoord(4,end)]);
    c3 = det([1 xcoord(1,1) xcoord(1,end); 1 xcoord(2,1) xcoord(2,end);...
        1 xcoord(4,1) xcoord(4,end)]);
    c4 = -det([1 xcoord(1,1) xcoord(1,end); 1 xcoord(2,1) xcoord(2,end);...
        1 xcoord(3,1) xcoord(3,end)]);
```

```
d1 = -det([1 xcoord(2,1:end-1); 1 xcoord(3,1:end-1);...
    1 xcoord(4,1:end-1)]);
d2 = det([1 xcoord(1,1:end-1); 1 xcoord(3,1:end-1);...
    1 xcoord(4,1:end-1)]);
d3 = -det([1 xcoord(1,1:end-1); 1 xcoord(2,1:end-1);...
    1 xcoord(4,1:end-1)]);
d4 = det([1 xcoord(1,1:end-1); 1 xcoord(2,1:end-1);...
    1 xcoord(3,1:end-1)]);
%
% INTEGRATION POINTS LOOP
%
for j1=1:nph
    chi = sg(3*(j1-1)+1);
    eta = sg(3*(j1-1)+2);
    tau = sg(3*j1);
    %
    % GEOMETRY APPROACH
    %
    SHPa(4) = tau;
    SHPa(3) = eta; SHPa(2) = chi; SHPa(1) = 1.-chi-eta-tau;
    chiG = 0.0; etaG = 0.0; tauG = 0.0;
    for k1=1:4
        chiG = chiG + SHPa(k1)*xcoord(k1,1);
        etaG = etaG + SHPa(k1)*xcoord(k1,2);
        tauG = tauG + SHPa(k1)*xcoord(k1,3);
    end
    %
    % SHAPE FUNCTION COMPUTATION
    %
    SHP(1) = (a1 + b1*chiG + c1*etaG + d1*tauG)/jcob;
    dSHPx(1) = b1/jcob; dSHPy(1) = c1/jcob; dSHPz(1) = d1/jcob;
    SHP(2) = (a2 + b2*chiG + c2*etaG + d2*tauG)/jcob;
    dSHPx(2) = b2/jcob; dSHPy(2) = c2/jcob; dSHPz(2) = d2/jcob;
    SHP(3) = (a3 + b3*chiG + c3*etaG + d3*tauG)/jcob;
    dSHPx(3) = b3/jcob; dSHPy(3) = c3/jcob; dSHPz(3) = d3/jcob;
    SHP(4) = (a4 + b4*chiG + c4*etaG + d4*tauG)/jcob;
    dSHPx(4) = b4/jcob; dSHPy(4) = c4/jcob; dSHPz(4) = d4/jcob;
    %
    % COMPUTE MATRICES OF SHAPE FUNCTION DERIVATIVES
    %
    B = [dSHPx(1) 0 0 dSHPx(2) 0 0 dSHPx(3) 0 0 dSHPx(4) 0 0; ...
        0 dSHPy(1) 0 0 dSHPy(2) 0 0 dSHPy(3) 0 0 dSHPy(4) 0;...
        0 0 dSHPz(1) 0 0 dSHPz(2) 0 0 dSHPz(3) 0 0 dSHPz(4);...
        dSHPy(1) dSHPx(1) 0 dSHPy(2) dSHPx(2) 0 dSHPy(3) dSHPx(3)...
        0 dSHPy(4) dSHPx(4) 0;...
        dSHPz(1) 0 dSHPx(1) dSHPz(2) 0 dSHPx(2) dSHPz(3) 0 dSHPx(3)...
        dSHPz(4) 0 dSHPx(4);...
        0 dSHPz(1) dSHPy(1) 0 dSHPz(2) dSHPy(2) 0 dSHPz(3) dSHPy(3)...
        0 dSHPz(4) dSHPy(4)];
    Ghat = [dSHPx(1) 0 0 dSHPx(2) 0 0 dSHPx(3) 0 0 dSHPx(4) 0 0; ...
        0 dSHPx(1) 0 0 dSHPx(2) 0 0 dSHPx(3) 0 0 dSHPx(4) 0;...
        0 0 dSHPx(1) 0 0 dSHPx(2) 0 0 dSHPx(3) 0 0 dSHPx(4);...
        dSHPy(1) 0 0 dSHPy(2) 0 0 dSHPy(3) 0 0 dSHPy(4) 0 0; ...
        0 dSHPy(1) 0 0 dSHPy(2) 0 0 dSHPy(3) 0 0 dSHPy(4) 0;...
        0 0 dSHPy(1) 0 0 dSHPy(2) 0 0 dSHPy(3) 0 0 dSHPy(4);...
        dSHPz(1) 0 0 dSHPz(2) 0 0 dSHPz(3) 0 0 dSHPz(4) 0 0; ...
        0 dSHPz(1) 0 0 dSHPz(2) 0 0 dSHPz(3) 0 0 dSHPz(4) 0;...
        0 0 dSHPz(1) 0 0 dSHPz(2) 0 0 dSHPz(3) 0 0 dSHPz(4)];

    if num_iter_prev == 0
        % Only stiffness matrix r1 is nonzero at the begining.
        r1(index,index) = r1(index,index) + (B'*D*B)*jcob*wg(j1);
    else
        for indI=NIc+1:num_iter_prev % summation over I.
            %
            % CALCULATE Ahat
            %
            AhatI=zeros(6,9);
            thetaI = Ghat*Fprev(index,indI);
            thetaxt=thetaI(1:3)';
            thetayt=thetaI(4:6)';
            thetazt=thetaI(7:9)';
```

```
    AhatI(1,1:3)=thetaxt;
    AhatI(2,4:6)=thetayt;
    AhatI(3,7:9)=thetazt;
    AhatI(4,1:6)=[thetayt thetaxt];
    AhatI(5,4:9)=[thetazt thetayt];
    AhatI(6,1:3)=thetazt;
    AhatI(6,7:9)=thetaxt;
    %
    % CALCULATE f1
    %
    XvecI = D*B*Fprev(index,indI);
    XI = zeros(3); %matrix expresion of XvecI.
    XI([1 5 9]) = XvecI([1 2 3]);
    XI(2) = XvecI(4); XI(4) = XI(2);
    XI(3) = XvecI(6); XI(7) = XI(3);
    XI(6) = XvecI(5); XI(8) = XI(6);
    f1I = [XI(1).*eye(3) XI(4).*eye(3) XI(7).*eye(3); ...
        XI(2).*eye(3) XI(5).*eye(3) XI(8).*eye(3);...
        XI(3).*eye(3) XI(6).*eye(3) XI(9).*eye(3)];
    %
    % CALCULATE r2I AND r3I
    %
    r2{indI}(index,index) = r2{indI}(index,index) + ...
        B'*D*AhatI*Ghat*jcob*wg(j1);
    r3{indI}(index,index) = r3{indI}(index,index) + ...
        Ghat'*f1I*Ghat*jcob*wg(j1);
    %
    % SECOND LOOP TO CALCULATE r4IJ AND r5IJ
    %
    for indJ=1:num_iter_prev % summation over J.
        %
        % CALCULATE Ahat
        %
        AhatJ=zeros(6,9);
        thetaJ = Ghat*Fprev(index,indJ);
        thetaxt=thetaJ(1:3)';
        thetayt=thetaJ(4:6)';
        thetazt=thetaJ(7:9)';
        AhatJ(1,1:3)=thetaxt;
        AhatJ(2,4:6)=thetayt;
        AhatJ(3,7:9)=thetazt;
        AhatJ(4,1:6)=[thetayt thetaxt];
        AhatJ(5,4:9)=[thetazt thetayt];
        AhatJ(6,1:3)=thetazt;
        AhatJ(6,7:9)=thetaxt;
        %
        % CALCULATE f1
        %
        XvecIJ = D*AhatI*thetaJ;
        XIJ = zeros(3);
        XIJ([1 5 9]) = XvecIJ([1 2 3]);
        XIJ(2) = XvecIJ(4); XIJ(4) = XIJ(2);
        XIJ(3) = XvecIJ(6); XIJ(7) = XIJ(3);
        XIJ(6) = XvecIJ(5); XIJ(8) = XIJ(6);
        f1IJ =[XIJ(1).*eye(3) XIJ(4).*eye(3) XIJ(7).*eye(3);...
            XIJ(2).*eye(3) XIJ(5).*eye(3) XIJ(8).*eye(3);...
            XIJ(3).*eye(3) XIJ(6).*eye(3) XIJ(9).*eye(3)];
        %
        % CALCULATE r4IJ and r5IJ
        %
        r4{(indI-1)*num_iter_prev+indJ}(index,index) = ...
            r4{(indI-1)*num_iter_prev+indJ}(index,index) + ...
            Ghat'*AhatI'*D*AhatJ*Ghat*jcob*wg(j1);
        r5{(indI-1)*num_iter_prev+indJ}(index,index) = ...
            r5{(indI-1)*num_iter_prev+indJ}(index,index) + ...
            Ghat'*f1IJ*Ghat*jcob*wg(j1);
    end
end
for indI=1:NIc
    %
    % CALCULATE Ahat
    %
```

```
                AhatI=zeros(6,9);
                thetaI = Ghat*Fprev(index,indI);
                thetaxt=thetaI(1:3)';
                thetayt=thetaI(4:6)';
                thetazt=thetaI(7:9)';
                AhatI(1,1:3)=thetaxt;
                AhatI(2,4:6)=thetayt;
                AhatI(3,7:9)=thetazt;
                AhatI(4,1:6)=[thetayt thetaxt];
                AhatI(5,4:9)=[thetazt thetayt];
                AhatI(6,1:3)=thetazt;
                AhatI(6,7:9)=thetaxt;
                for indJ=NIc+1:num_iter_prev
                    %
                    % CALCULATE Ahat
                    %
                    AhatJ=zeros(6,9);
                    thetaJ = Ghat*Fprev(index,indJ);
                    thetaxt=thetaJ(1:3)';
                    thetayt=thetaJ(4:6)';
                    thetazt=thetaJ(7:9)';
                    AhatJ(1,1:3)=thetaxt;
                    AhatJ(2,4:6)=thetayt;
                    AhatJ(3,7:9)=thetazt;
                    AhatJ(4,1:6)=[thetayt thetaxt];
                    AhatJ(5,4:9)=[thetazt thetayt];
                    AhatJ(6,1:3)=thetazt;
                    AhatJ(6,7:9)=thetaxt;
                    %
                    % CALCULATE f1
                    %
                    XvecIJ = D*AhatI*thetaJ;
                    XIJ = zeros(3);
                    XIJ([1 5 9]) = XvecIJ([1 2 3]);
                    XIJ(2) = XvecIJ(4); XIJ(4) = XIJ(2);
                    XIJ(3) = XvecIJ(6); XIJ(7) = XIJ(3);
                    XIJ(6) = XvecIJ(5); XIJ(8) = XIJ(6);
                    f1IJ =[XIJ(1).*eye(3) XIJ(4).*eye(3) XIJ(7).*eye(3);...
                        XIJ(2).*eye(3) XIJ(5).*eye(3) XIJ(8).*eye(3);...
                        XIJ(3).*eye(3) XIJ(6).*eye(3) XIJ(9).*eye(3)];
                    %
                    % CALCULATE r4IJ and r5IJ
                    %
                    r4{(indI-1)*num_iter_prev+indJ}(index,index) = ...
                        r4{(indI-1)*num_iter_prev+indJ}(index,index) + ...
                        Ghat'*AhatI'*D*AhatJ*Ghat*jcob*wg(j1);
                    r5{(indI-1)*num_iter_prev+indJ}(index,index) = ...
                        r5{(indI-1)*num_iter_prev+indJ}(index,index) + ...
                        Ghat'*f1IJ*Ghat*jcob*wg(j1);
                end
            end
        end
    end
end
return
```

Soon after the `fem3DHyperelastic` routine, the code calls the `ElemstiffHyperelastic` routine, which computes the s_1, s_{2I} and s_{3IJ} matrices, see Terms $T2$, $T4$, …

In principle, coordinate s is defined over the boundary of the solid, which may be triangulated and discretized with the help of linear triangular elements, for instance. In this case, however, our implementation looks for the nearest node to a particular s coordinate to apply the load. This means that only nodal forces are considered, for simplicity. In turn, the s coordinate can be parameterized in the form of a one-dimensional array of nodes.

```
function elemstiffHyperelastic(coor)
% function elemstiff(coor)
% Computes s1, s2I and s3IJ matrices
% Universidad de Zaragoza - 2015
global s1 s2 s3
global num_iter_prev Gprev

nen = numel(coor);

% As in fem3DHyperelastic only new matrices s2I and s3IJ are calculated.
% s1 is calculated only the first time.
NIc = 0;
if num_iter_prev > 0
    NIc = size(s2,1); % Number of matrices s2 already calculated.
    if isempty(s2)==1, NIc = 0; end
end
if NIc > 0
    % Save previously calculated matrices on auxiliary variables.
    s2_aux = s2; s3_aux = s3;
end
%
% ALLOCATE MEMORY AND INITIALIZE
%
if num_iter_prev == 0
    s1 = sparse(nen,nen);
end
s2 = cell(num_iter_prev,1);
s3 = cell(num_iter_prev*num_iter_prev,1);
for i1=1:num_iter_prev, s2{i1} = sparse(nen,nen); end
for i1=1:num_iter_prev*num_iter_prev
    s3{i1} = sparse(nen,nen);
end
%
% ADD ALREADY CALCULATED MATRICES
%
if NIc>0
    s2(1:NIc) = s2_aux;
    Incre = num_iter_prev-NIc;
    if Incre>0
        IndMat = zeros(NIc*NIc,NIc); IndMat(:,1) = [1:NIc*NIc]';
        for i1=2:NIc
            IndMat(:,i1) = IndMat(:,i1-1) + Incre.*ones(size(IndMat,1),1);
        end
        IndMat = reshape(IndMat(:),[NIc,NIc*NIc]);
        IM = IndMat(:,1:NIc+1:end);
        s3(IM(:)) = s3_aux;
    else
        s3 = s3_aux;
    end
end
%
% CALCULATE NEW MATRICES
%
X = coor(1:nen-1)'; Y = coor(2:nen)'; % Coordinates of elements.
L = Y - X; % Longitude of each element for parametriced variable.
sg = [-1.0/sqrt(3.0) 1.0/sqrt(3.0)]; wg = ones(2,1); % Gauss points.
npg = numel(sg);
```

```
for i1=1:nen-1
    c = zeros(1,npg);
    N = zeros(nen,npg);
    c(1,:) = 0.5.*(1.0-sg).*X(i1) + 0.5.*(1.0+sg).*Y(i1);
    N(i1+1,:) = (c(1,:)-X(i1))./L(i1);
    N(i1,:) = (Y(i1)-c(1,:))./L(i1);
    for j1=1:npg
        if num_iter_prev == 0
            s1 = s1 + N(:,j1)*N(:,j1)'*0.5.*wg(j1).*L(i1);
        else
            for indI=NIc+1:num_iter_prev
                s2{indI}(:,:) = s2{indI}(:,:) + ...
         (N(:,j1)'*Gprev(:,indI))*N(:,j1)*N(:,j1)'*0.5.*wg(j1).*L(i1);
                for indJ=1:num_iter_prev
                    s3{(indI-1)*num_iter_prev+indJ}(:,:) = ...
                        s3{(indI-1)*num_iter_prev+indJ}(:,:) + ...
         (N(:,j1)'*Gprev(:,indI))*(N(:,j1)'*Gprev(:,indJ))...
         *N(:,j1)*N(:,j1)'*0.5.*wg(j1).*L(i1);
                end
            end
            for indI=1:NIc
                for indJ=NIc+1:num_iter_prev
                    s3{(indI-1)*num_iter_prev+indJ}(:,:) = ...
                        s3{(indI-1)*num_iter_prev+indJ}(:,:) + ...
         (N(:,j1)'*Gprev(:,indI))*(N(:,j1)'*Gprev(:,indJ))...
         *N(:,j1)*N(:,j1)'*0.5.*wg(j1).*L(i1);
                end
            end
        end
    end
end
return
```

Finally, the program calls for the `enrichment` routine, responsible for the computation of the new terms R and S, see Eq. (4.7), in the approximation.

```
function  [R,S,iter] = enrichment(S0,F,G,TOL,IndBcnode,num_iter)
% function  [R,S,iter] = enrichment(S0,F,G,TOL,IndBcnode,num_iter)
% Computes a new sumand by fixed-point algorithm using PGD
% Universidad de Zaragoza - 2015
global vx vp
global num_iter_prev
global r1 r2 r3 r4 r5 s1 s2 s3

R = zeros(size(F,1),1); R0 = R; % Initial value R to compare in first loop.
h = size(vx,2); % Number functions of the source.
ExitFlag = 1;
iter = 0;
mxit = 100; % #ăof possible iterations for the fixed point algorithm.
Free = setdiff(1:numel(F(:,1)),IndBcnode);
%
% FIXED POINT ALGORITHM
%
while ExitFlag > TOL
    %
    % LOOKING FOR R, KNOWNING S
    %
    matrixR = r1*(S0'*s1*S0); % T2K
    sourceR = zeros(size(F,1),1);
    for k1=1:h
        sourceR = sourceR + vx(:,k1)*(S0'*vp(:,k1)); % source term.
    end
    for i1=1:num_iter-1
        sourceR = sourceR - (r1*F(:,i1))*(S0'*s1*G(:,i1)); % T2F
    end
    for i1=1:num_iter_prev
        matrixR = matrixR + ...
            r2{i1}.*(S0'*s2{i1}*S0) + ...
            r2{i1}'.*(S0'*s2{i1}*S0) + ...
            r3{i1}.*(S0'*s2{i1}*S0); % T4K, T6K, T9K
        for j1=1:num_iter-1
```

```
            sourceR = sourceR - ...
                r2{i1}*F(:,j1)*(S0'*s2{i1}*G(:,j1)) - ...
                r2{i1}'*F(:,j1)*(S0'*s2{i1}*G(:,j1)) - ...
                r3{i1}*F(:,j1)*(S0'*s2{i1}*G(:,j1)); % T4F, T6F, T9F
        end
    end
    for i1=1:num_iter_prev*num_iter_prev
        matrixR = matrixR + ...
            r4{i1}.*(S0'*s3{i1}*S0) + ...
            0.5*r5{i1}.*(S0'*s3{i1}*S0); % T8K, T10K
        for j1=1:num_iter-1
            sourceR = sourceR - ...
                r4{i1}*F(:,j1).*(S0'*s3{i1}*G(:,j1)) - ...
                0.5*r5{i1}*F(:,j1).*(S0'*s3{i1}*G(:,j1)); % T8F, T10F
        end
    end
    %
    % SOLVE R
    %
    R(Free) = matrixR(Free,Free)\sourceR(Free);
    %
    % LOOKING FOR S, KNOWNING R
    %
    matrixS = (R'*r1*R).*s1; % T2K
    sourceS = zeros(size(G,1),1);
    for k1=1:h
        sourceS = sourceS + (R'*vx(:,k1))*vp(:,k1); % source term.
    end
    for i1=1:num_iter-1
        sourceS = sourceS - (R'*r1*F(:,i1))*(s1*G(:,i1)); % T2F
    end
    for i1=1:num_iter_prev
        matrixS = matrixS + ...
            (R'*r2{i1}*R).*s2{i1} + ...
            (R'*r2{i1}'*R).*s2{i1} + ...
            (R'*r3{i1}*R).*s2{i1}; % T4K, T6K, T9K
        for j1=1:num_iter-1
            sourceS = sourceS - ...
                (R'*r2{i1}*F(:,j1))*s2{i1}*G(:,j1) - ...
                (R'*r2{i1}'*F(:,j1))*s2{i1}*G(:,j1) - ...
                (R'*r3{i1}*F(:,j1))*s2{i1}*G(:,j1); % T4F, T6F, T9F
        end
    end
    for i1=1:num_iter_prev*num_iter_prev
        matrixS = matrixS + ...
            (R'*r4{i1}*R).*s3{i1} + ...
            0.5*(R'*r5{i1}*R).*s3{i1}; % T8K, T10K
        for j1=1:num_iter-1
            sourceS = sourceS - ...
                (R'*r4{i1}*F(:,j1)).*(s3{i1}*G(:,j1)) - ...
                0.5*(R'*r5{i1}*F(:,j1)).*(s3{i1}*G(:,j1)); % T8F, T10F
        end
    end
    %
    % SOLVE S
    %
    S = matrixS\sourceS;
    S = S./norm(S); % We normalize S. R takes care of alpha constant.
    %
    % COMPUTING STOP CRITERIA
    %
    error = max(abs(sum(R0-R)),abs(sum(S0-S)));
    R0 = R; S0 = S;
    iter = iter + 1;
    if iter>mxit || abs(error)<TOL,
        return
    end
end

return
```

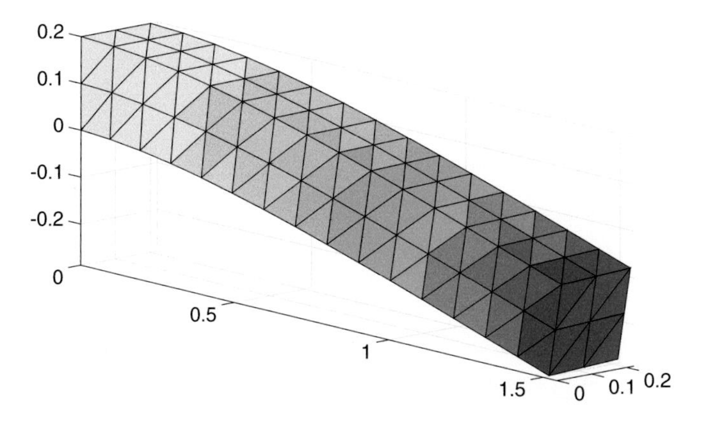

Fig. 4.2 Deformed configuration of the cantilever beam at the end of the simulation

After execution, whose most important part is devoted to the computation of the off-line terms in the approximation, the program depicts the deformed configuration of the beam, see Fig. 4.2.

Chapter 5
PGD for Dynamical Problems

Simulations are like Michelin star restaurants but should be like McDonalds: ubiquitous and standardised.

—Craig McIlhenny

Abstract This chapter develops the application of PGD methods initial and boundary value problems, with an eye towards the (non-linear) solid dynamics equations.

Model order reduction of initial and boundary value problems (IBVP) is a particularly challenging task. In this chapter we explain some interesting concepts related mostly with solid dynamics, taken as model problem to this end.

In [41] a method was developed that takes the field of initial conditions as a parameter to develop a very efficient dynamical integrator. However, the field of initial conditions (displacement, velocity) is in fact a parameter of infinite dimension, and hence hard to parameterize adequately. In this chapter we analyze how to do it in a proper way so as to render a very fast method, amenable for real-time simulation, even under very astringent conditions. Other approaches to the problem, such as a space-time one, can be found at [18], for instance.

5.1 Taking Initial Conditions as Parameters

As mentioned before, in [41] a method is developed based on PGD that acts as a sort of *black box* integrator in time. Given the converged displacement and velocity field of the solid at time step t, $\boldsymbol{u}^t$ and $\dot{\boldsymbol{u}}^t$, respectively, as parameters, the method returns the displacement and velocity fields at time $t + \triangle t$, see Fig. 5.1.

Once semi-discretized in space, the displacement and velocity fields are no longer of infinite dimension, but usual engineering finite element meshes involve tens of

E. Cueto et al., *Proper Generalized Decompositions*, SpringerBriefs in Applied Sciences and Technology, DOI 10.1007/978-3-319-29994-5_5

Fig. 5.1 Sketch of the proposed method for the integration of solid dynamics in the PGD framework. Converged displacement and velocity fields at time step t are taken as parameters, so as to provide, without the need of any matrix inversion, the displacement and velocity fields at time step $t + \Delta t$

thousands to millions of degrees of freedom. This would imply to have into account millions of parameters, something out of reach even for PGD methods.

In order to avoid this enormous number of parameters, in [41] the use of Proper Orthogonal Decomposition [43, 48, 49] methods so as to employ a minimal number of parameters is proposed. In this way, initial displacement and velocity field can be optimally parameterized with a minimal number of degrees of freedom. The price to pay is to project the results of the integration at time $t + \Delta t$ onto the POD basis so as to be taken as parameters (initial conditions) for a subsequent integration to obtain $\boldsymbol{u}^{t+2\Delta t}$ and $\dot{\boldsymbol{u}}^{t+2\Delta t}$.

We provide details of the variational formulation in the subsequent sections.

5.2 Developing the Weak Form of the Problem

We consider the general problem of solid dynamics, in which we look for the displacement field,

$$\boldsymbol{u} : \bar{\Omega} \times]0, T] \times \mathcal{I} \times \mathcal{J} \rightarrow \mathbb{R}^3,$$

where $\mathcal{I} = [\boldsymbol{u}_0^-, \boldsymbol{u}_0^+]$ and $\mathcal{J} = [\dot{\boldsymbol{u}}_0^-, \dot{\boldsymbol{u}}_0^+]$ represent the considered intervals of variation of initial boundary conditions, $\boldsymbol{u}_0$ and $\dot{\boldsymbol{u}}_0$, taken as parameters. To obtain a parametric solution for any initial condition (within these intervals), it is therefore necessary to define a new (triply-) weak form:

given $\boldsymbol{f}$, $\boldsymbol{g}$, $\boldsymbol{h}$, $\boldsymbol{u}_0$ *and* $\dot{\boldsymbol{u}}_0$ *find* $\boldsymbol{u}(t) \in \mathcal{S}_t = \{\boldsymbol{u}|\boldsymbol{u}(\boldsymbol{x}, t) = \boldsymbol{g}(\boldsymbol{x}, t),\ \boldsymbol{x} \in \Gamma_u,\ \boldsymbol{u} \in \mathcal{H}^1(\Omega)\}$, $t \in [0, T]$, *such that for all* $\boldsymbol{u}^* in \mathcal{V}\{\boldsymbol{u}|\boldsymbol{u}(\boldsymbol{x}, t) = \boldsymbol{0},\ \boldsymbol{x} \in \Gamma_u,\ \boldsymbol{u} \in \mathcal{H}^1(\Omega)\}$,

$$(\boldsymbol{u}^*, \rho\ddot{\boldsymbol{u}}) + a(\boldsymbol{u}^*, \boldsymbol{u}) = (\boldsymbol{u}^*, \boldsymbol{f}) + (\boldsymbol{u}^*, \boldsymbol{h})_\Gamma \tag{5.1a}$$

$$(\boldsymbol{u}^*, \rho\boldsymbol{u}(0)) = (\boldsymbol{u}^*, \rho\boldsymbol{u}_0) \tag{5.1b}$$

$$(\boldsymbol{u}^*, \rho\dot{\boldsymbol{u}}(0)) = (\boldsymbol{u}^*, \rho\dot{\boldsymbol{u}}_0), \tag{5.1c}$$

where:

$$a(\boldsymbol{u}^*, \boldsymbol{u}) = \int_{\mathcal{I}} \int_{\mathcal{J}} \int_{\Omega} \nabla^s \boldsymbol{u}^* : \mathbf{C} : \nabla^s \boldsymbol{u} \; d\Omega d\dot{\boldsymbol{u}}_0 d\boldsymbol{u}_0,$$

$$(\boldsymbol{u}^*, \boldsymbol{f}) = \int_{\mathcal{I}} \int_{\mathcal{J}} \int_{\Omega} \boldsymbol{u}^* \boldsymbol{f} \; d\Omega d\dot{\boldsymbol{u}}_0 d\boldsymbol{u}_0,$$

$$(\boldsymbol{u}^*, \boldsymbol{h})_{\Gamma} = \int_{\mathcal{I}} \int_{\mathcal{J}} \int_{\Gamma_t} \boldsymbol{u}^* \boldsymbol{h} \; d\Gamma d\dot{\boldsymbol{u}}_0 d\boldsymbol{u}_0.$$

The next step is, by means of appropriate finite-dimensional approximations to $\mathcal{S}_t$ and $\mathcal{V}$, $\mathcal{S}_t^h$ and $\mathcal{V}^h$, respectively, to semi-discretize the weak form so as to obtain the following problem:

given $\boldsymbol{f}, \boldsymbol{g}, \boldsymbol{h}, \boldsymbol{u}_0$ *and* $\dot{\boldsymbol{u}}_0$ *find* $\boldsymbol{u}^h(t) = \boldsymbol{v}^h + \boldsymbol{g}^h \in \mathcal{S}_t^h$ *(note that* $\boldsymbol{g}(\boldsymbol{x}, t) = \boldsymbol{u}(\boldsymbol{x}, t)$ *on* Γ_u*) such that for every* $\boldsymbol{u}^{*h} \in \mathcal{V}^h$,

$$(\boldsymbol{u}^{*h}, \rho \ddot{\boldsymbol{v}}^h) + a(\boldsymbol{u}^{*h}, \boldsymbol{v}) = (\boldsymbol{u}^{*h}, \boldsymbol{f}) + (\boldsymbol{u}^{*h}, \boldsymbol{h})_{\Gamma} - (\boldsymbol{u}^{*h}, \rho \ddot{\boldsymbol{g}}^h) - a(\boldsymbol{u}^{*h}, \boldsymbol{g}^h), \tag{5.2a}$$

$$(\boldsymbol{u}^{*h}, \rho \boldsymbol{v}^h(0)) = (\boldsymbol{u}^{*h}, \rho \boldsymbol{u}_0) - (\boldsymbol{u}^{*h}, \rho \boldsymbol{g}^h(0)), \tag{5.2b}$$

$$(\boldsymbol{u}^{*h}, \rho \dot{\boldsymbol{v}}^h(0)) = (\boldsymbol{u}^{*h}, \rho \dot{\boldsymbol{u}}_0) - (\boldsymbol{u}^{*h}, \rho \dot{\boldsymbol{g}}^h(0)). \tag{5.2c}$$

This provides a sort of response surface or parametric solution (thus the name *computational vademecum* coined in [26]) to the problem (5.1) for *any* initial conditions.

If we consider direct integration in time (remember that we look for an interactive method, so that this prevents us from using a space-time approach) the sought displacement field will be approximated in a PGD framework as a finite series of separable functions,

$$\boldsymbol{v}^h(\boldsymbol{x}, t, \boldsymbol{u}_0, \dot{\boldsymbol{u}}_0) = \left[\sum_{i=1}^{n} \boldsymbol{F}^i(\boldsymbol{x}) \circ \boldsymbol{G}^i(\boldsymbol{u}_0) \circ \boldsymbol{H}^i(\dot{\boldsymbol{u}}_0) \right] \circ \boldsymbol{d}(t), \tag{5.3}$$

where the nodal coefficients $\boldsymbol{d}(t)$ carry out all the time-dependency of the solution and the symbol "$\circ$" stands for the entry-wise Hadamard or Schur multiplication of vectors.

Functions $\boldsymbol{F}, \boldsymbol{G}$ and $\boldsymbol{H}$ will be expressed in terms of low (here, three-) dimensional finite element basis functions. As usual, these are computed by means of a greedy algorithm in which one sum is computed at a time, while one product is computed in a fixed point, alternated directions algorithm. Thus, having an approximation to $\boldsymbol{v}^h$ converged at iteration n, the $(n+1)$-th term is obtained as

$$\boldsymbol{v}^{n+1}(\boldsymbol{x}, t, \boldsymbol{u}_0, \dot{\boldsymbol{u}}_0) = \left[\sum_{i=1}^{n} \boldsymbol{F}^i(\boldsymbol{x}) \circ \boldsymbol{G}^i(\boldsymbol{u}_0) \circ \boldsymbol{H}^i(\dot{\boldsymbol{u}}_0) + \boldsymbol{R}(\boldsymbol{x}) \circ \boldsymbol{S}(\boldsymbol{u}_0) \circ \boldsymbol{T}(\dot{\boldsymbol{u}}_0) \right] \circ \boldsymbol{d}(t).$$

By substituting the approximations to $\boldsymbol{v}^h$ and $\boldsymbol{w}^h$ into the weak form of the problem, Eq. (5.2), we arrive at a semi-discrete problem. One of the most salient features of this method relies in its ability of advancing in time using any time integrator existing in the literature (particularly, energy and momentum conserving schemes.) Of course, any other parametric dependence, such as the one on the position of the applied load, see Chap. 3, can be considered at the same time.

Still one ingredient of the implementation is missing. Instead of considering the whole time interval $]0, T]$ we look for the solution within a generic time increment $]0, \Delta t]$:

$$\boldsymbol{v} : \bar{\Omega} \times]0, \Delta t] \times \mathcal{I} \times \mathcal{J} \times [h^-, h^+] \rightarrow \mathbb{R}^3,$$

where Δt represents the necessary time to response prescribed by the particular envisaged application. For instance, for haptic feedback applications, a physically realistic sensation of touch needs for some 500 Hz to 1 kHz feedback rate. This implies to take Δt on the order of 0.001 seconds. This value Δt could be smaller or greater than the necessary time step amplitude needed for stability of the chosen time integrator.

5.3 Matrix Form of the Problem

5.3.1 Time Integration of the Equations of Motion

As usual, we start from the weak form of the solid dynamic equations, Eq. (5.1), i.e., finding the displacement $\boldsymbol{u} \in \mathcal{H}^1$ such that for all $\boldsymbol{u}^* \in \mathcal{H}_0^1$:

$$\int_\Omega \boldsymbol{u}^* \rho \ddot{\boldsymbol{u}} d\Omega + \int_\Omega \nabla_s \boldsymbol{u}^* : \mathsf{C} : \nabla_s \boldsymbol{u} d\Omega = \int_{\Gamma_{t2}} \boldsymbol{u}^* \cdot \boldsymbol{t} d\Gamma. \tag{5.4}$$

Once semi-discretized in space, we can identify a term where a mass matrix appears,

$$\boldsymbol{M}_m = \int_\Omega \boldsymbol{u}^{*h} \rho \ddot{\boldsymbol{u}}^h d\Omega,$$

and a term usually identified as the stiffness matrix

$$\boldsymbol{K}_m = \int_\Omega \nabla_s \boldsymbol{u}^{*h} : \mathsf{C} : \nabla_s \boldsymbol{u}^h d\Omega.$$

In the sequel we omit, if no risk of confusion exists, the superscript h, so that all vectors represent the set of nodal unknowns for the problem. For the integration in time of these equations we have several options. In general you can employ your favorite integration scheme. Here we are considering, both for its simplicity and good results, an energy-momentum conserving scheme developed in [15]. This scheme has two sub-steps which compute a predictor of the displacement vector at time step $u_{t+(\Delta t/2)}$ in the first one and subsequently a correction $u_{t+\Delta t}$ in the second sub-step.

The first sub-step has the following form:

$$\boldsymbol{M}_m \ddot{\boldsymbol{u}}_{t+(\Delta t/2)} + \boldsymbol{K}_m \boldsymbol{u}_{t+(\Delta t/2)} = \boldsymbol{F}_{t+(\Delta t/2)}.$$

Employing classical finite difference approaches for the time derivatives,

$$\ddot{\boldsymbol{u}}_{t+(\Delta t/2)} = \frac{\dot{\boldsymbol{u}}_{t+(\Delta t/2)} - \dot{\boldsymbol{u}}_t}{\Delta t/4} - \ddot{\boldsymbol{u}}_t,$$

$$\dot{\boldsymbol{u}}_{t+(\Delta t/2)} = \frac{\boldsymbol{u}_{t+(\Delta t/2)} - \boldsymbol{u}_t}{\Delta t/4} - \dot{\boldsymbol{u}}_t.$$

Applying these expressions to the first sub-step, after some simple algebra, we obtain the final expression for the sub-step 1:

$$\left[\left[\frac{16}{\Delta t^2}\right]\boldsymbol{M}_m + \boldsymbol{K}_m\right]\cdot \boldsymbol{u}_{t+(\Delta t/2)} = \boldsymbol{F}_{t+(\Delta t/2)} + \left[\frac{16}{\Delta t^2}\right]\boldsymbol{M}_m \cdot \boldsymbol{u}_t + \left[\frac{8}{\Delta t}\right]\boldsymbol{M}_m \cdot \dot{\boldsymbol{u}}_t + \boldsymbol{M}_m \cdot \ddot{\boldsymbol{u}}_t. \quad (5.5)$$

The second sub-step has the following form:

$$\boldsymbol{M}_m \ddot{\boldsymbol{u}}_{t+\Delta t} + \boldsymbol{K}_m \boldsymbol{u}_{t+\Delta t} = \boldsymbol{F}_{t+\Delta t}.$$

Again, by employing classical finite differences for the time derivatives,

$$\ddot{\boldsymbol{u}}_{t+\Delta t} = \frac{\dot{\boldsymbol{u}}_t}{\Delta t} - \left[\frac{4}{\Delta t}\right]\dot{\boldsymbol{u}}_{t+(\Delta t/2)} + \left[\frac{3}{\Delta t}\right]\dot{\boldsymbol{u}}_{t+\Delta t},$$

$$\dot{\boldsymbol{u}}_{t+\Delta t} = \frac{\boldsymbol{u}_t}{\Delta t} - \left[\frac{4}{\Delta t}\right]\boldsymbol{u}_{t+(\Delta t/2)} + \left[\frac{3}{\Delta t}\right]\boldsymbol{u}_{t+\Delta t}.$$

By substituting these expressions in the second sub-step, the final formula for the sub-step 2 that the reader can find in the code included in Sect. 5.4 is:

$$\left[\left[\frac{9}{\Delta t^2}\right]\boldsymbol{M}_m + \boldsymbol{K}_m\right]\cdot \boldsymbol{u}_{t+\Delta t} = \boldsymbol{F}_{t+\Delta t} - \left[\frac{19}{\Delta t^2}\right]\boldsymbol{M}_m \cdot \boldsymbol{u}_t - \left[\frac{5}{\Delta t}\right]\boldsymbol{M}_m \cdot \dot{\boldsymbol{u}}_t + \left[\frac{28}{\Delta t^2}\right]\boldsymbol{M}_m \cdot \boldsymbol{u}_{t+(\Delta t/2)}. \quad (5.6)$$

The strategy depicted in the previous section, when applied to the just explained time integration scheme, implies the construction of a PGD time integrator able to provide the value of $\boldsymbol{u}_{t+\triangle t}$ for any value of $\boldsymbol{u}_t$. In that framework, $\boldsymbol{u}_t$ acts in fact as a parameter. But recall that $\boldsymbol{u}_t$ represents the vector of nodal displacements at time step t. Therefore, it can consist of several millions of degrees of freedom, something our of reach even for PGD strategies!

In order to develop a suitable strategy, it is therefore of utmost importance to adequately parameterize the field of initial displacements at the beginning of the time step. In [41] this is done by employing a reduced-order basis instead of the traditional finite element one. And to do it by means of Proper Orthogonal Decompositions. This is explained in detail in what follows.

5.3.2 Computing a Reduced-Order Basis for the Field of Initial Conditions

For the sake of completeness, we briefly review here the basics of the POD technique for the computation of a reduced-order basis for the initial displacement field of the problem. Let us first assume that we have a collection of *snapshots*, i.e., finite element results for problems similar to the one at hand. By *similar* we mean results probably for the same solid, but possibly under different conditions, applied loads, boundary conditions, … We then store these snapshots column-wise in a matrix $\boldsymbol{Q}$ (more details can be found, for instance, in [54]). The next step is the computation of the so-called *auto-correlation matrix*,

$$\boldsymbol{c} = \boldsymbol{Q}\ \boldsymbol{Q}^T. \tag{5.7}$$

It can then be demonstrated that the best possible basis (that capturing the most of the energy of the system with the minimal number of degrees of freedom) is formed by the eigenvectors $\boldsymbol{\phi}$ of the problem

$$\boldsymbol{c}\ \boldsymbol{\phi} = \alpha \boldsymbol{\phi}.$$

By storing the nodal values (we assume to have N nodes in the mesh of the model) of the eigenvectors with the m biggest eigenvalues in a matrix

$$\mathsf{B} = \begin{pmatrix} \phi^1(\boldsymbol{x}_1) & \phi^2(\boldsymbol{x}_1) & \cdots & \phi^m(\boldsymbol{x}_1) \\ \phi^1(\boldsymbol{x}_2) & \phi^2(\boldsymbol{x}_2) & \cdots & \phi^m(\boldsymbol{x}_2) \\ \vdots & \vdots & \ddots & \vdots \\ \phi^1(\boldsymbol{x}_N) & \phi^2(\boldsymbol{x}_N) & \cdots & \phi^m(\boldsymbol{x}_N) \end{pmatrix}$$

we can therefore project the initial system of equations onto a reduced-order one by simply doing the change of variable

$$\boldsymbol{u}_t \approx \sum_{i=1}^{i=\mathtt{nrb}} \zeta_t^i \boldsymbol{\phi}^i = \boldsymbol{B}\,\boldsymbol{\zeta}_t,$$

so that we will finally face a system of $3m$ equations for $\boldsymbol{\zeta}_t$ instead of the original $3N$ for $\boldsymbol{u}_t$. The advantage of this strategy is that usually the number of reduced basis, $\mathtt{rnb} \ll N$ and therefore the resulting system of equations is generally much smaller.

5.3.3 Projection of the Equations onto a Reduced, Parametric Basis

For each sub-step within the time integration scheme we compute the PGD approximation to the solution $\boldsymbol{u}_{t+(\Delta t/2)}$ and $\boldsymbol{u}_{t+\Delta t}$ such that,

$$\begin{aligned}&\boldsymbol{u}_{t+(\Delta t/2)}(\boldsymbol{x}, \boldsymbol{\zeta}_t, \dot{\boldsymbol{\zeta}}_t, \ddot{\boldsymbol{\zeta}}_t, p_a, \boldsymbol{s})\\&= \sum_{i=1}^{n} \boldsymbol{N}^T(\boldsymbol{x})\boldsymbol{F}^i_{\boldsymbol{x}} \cdot \boldsymbol{N}^T(\boldsymbol{\zeta}_t)\boldsymbol{F}^i_{\boldsymbol{\zeta}_t} \cdot \boldsymbol{N}^T(\dot{\boldsymbol{\zeta}}_t)\boldsymbol{F}^i_{\dot{\boldsymbol{\zeta}}_t} \cdot \boldsymbol{N}^T(\ddot{\boldsymbol{\zeta}}_t)\boldsymbol{F}^i_{\ddot{\boldsymbol{\zeta}}_t} \cdot \boldsymbol{N}^T(p_a)\boldsymbol{F}^i_{p_a} \cdot \boldsymbol{N}^T(\boldsymbol{s})\boldsymbol{F}^i_{\boldsymbol{s}}\end{aligned} \tag{5.8}$$

$$\begin{aligned}&\boldsymbol{u}_{t+\Delta t}(\boldsymbol{x}, \boldsymbol{\zeta}_t, \dot{\boldsymbol{\zeta}}_t, \boldsymbol{\zeta}_{t+(\Delta t/2)}, p_a, \boldsymbol{s})\\&= \sum_{i=1}^{n} \boldsymbol{N}^T(\boldsymbol{x})\boldsymbol{F}^i_{\boldsymbol{x}} \cdot \boldsymbol{N}^T(\boldsymbol{\zeta}_t)\boldsymbol{F}^i_{\boldsymbol{\zeta}_t} \cdot \boldsymbol{N}^T(\dot{\boldsymbol{\zeta}}_t)\boldsymbol{F}^i_{\dot{\boldsymbol{\zeta}}_t} \cdot \boldsymbol{N}^T(\boldsymbol{\zeta}_{t+(\Delta t/2)})\boldsymbol{F}^i_{\boldsymbol{\zeta}_{t+(\Delta t/2)}} \cdot \boldsymbol{N}^T(p_a)\boldsymbol{F}^i_{p_a} \cdot \boldsymbol{N}^T(\boldsymbol{s})\boldsymbol{F}^i_{\boldsymbol{s}}\end{aligned} \tag{5.9}$$

where $\boldsymbol{x}$ represents the physical space, $\boldsymbol{u}_t$ is the vector of displacement degrees of freedom at time step t, $\dot{\boldsymbol{u}}_n$ is the vector of velocity degrees of freedom at time step t, $\ddot{\boldsymbol{u}}_n$ is the vector of nodal accelerations at time step t, $\boldsymbol{u}_{t+(\Delta t/2)}$ is the vector of nodal displacements at time step $t + (\Delta t/2)$, p_a is the classical loading parameter, its value varying continuously in the interval [0, 1]. It allows us to apply or not a load at a particular time step or to apply a ramp load, for instance. Finally, $\boldsymbol{s}$ represents the position of the load, as in Chap. 3.

We denote, as in the rest of the book, by $\boldsymbol{N}(\cdot)$ the vector of finite element shape functions employed to discretize the different dimensions of the problem. Note that we are considering a solution depending on the physical space $\boldsymbol{x}$ plus a number of parameters $\boldsymbol{\zeta} = [\zeta_1, \zeta_2, \ldots, \zeta_m]$, which in this case coincide with the chosen POD degrees of freedom parameterizing the fields of initial displacements and velocities. Taking also into account that the density parameter ρ and the symmetric gradients ∇_s in Eq. (5.1) depend solely on space coordinates, we can write the mass matrix, and the stiffness matrix of the problem in separated form as

$$M_m = \left[\int_{\Omega_x} N^T(x)\rho N(x) d\Omega_x\right] \cdot \left[\int_{\Omega_{\zeta_1}} N^T(\zeta_1) N(\zeta_1) d\Omega_{\zeta_1}\right] \cdots$$
$$\cdot \left[\int_{\Omega_{\zeta_m}} N^T(\zeta_m) N(\zeta_m) d\Omega_{\zeta_m}\right] \cdot \left[\int_{\Omega_{p_a}} N^T(p_a) N(p_a) d\Omega_{p_a}\right] \cdot \left[\int_{\Omega_s} N^T(s) N(s) d\Omega_s\right],$$

$$K_m = \left[\int_{\Omega_x} \nabla_s N^T(x) \mathbf{C} \nabla_s N(x) d\Omega_x\right] \cdot \left[\int_{\Omega_{\zeta_1}} N^T(\zeta_1) N(\zeta_1) d\Omega_{\zeta_1}\right] \cdots$$
$$\cdot \left[\int_{\Omega_{\zeta_m}} N^T(\zeta_m) N(\zeta_m) d\Omega_{\zeta_m}\right] \left[\int_{\Omega_{p_a}} N^T(p_a) N(p_a) d\Omega_{p_a}\right] \cdot \left[\int_{\Omega_s} N^T(s) N(s) d\Omega_s\right].$$

The influence on the solution of the number of parameters m (terms in the POD basis of the initial conditions) chosen to parameterize the fields of initial conditions was deeply analyzed in [41].

The PGD final solution uses the PGD solutions for each sub-step, Eqs. (5.5) and (5.6). Starting from $\boldsymbol{u}_0$, $\dot{\boldsymbol{u}}_0$ and $\ddot{\boldsymbol{u}}_0$ (in fact, their projection onto the POD basis!), by using the first sub-step vademecum we obtain $\boldsymbol{u}_{0+1/2}$. These values are then introduced in second sub-step vademecum which in turn uses as input parameters $\boldsymbol{u}_0$, $\dot{\boldsymbol{u}}_0$ and $\boldsymbol{u}_{0+1/2}$ (computed in the first sub-step). The value returned by the second vademecum is $\boldsymbol{u}_1$. We then change the time step, and apply the new input parameters $\boldsymbol{u}_1$, $\dot{\boldsymbol{u}}_1$ and $\ddot{\boldsymbol{u}}_1$ in the first vademecum to obtain $\boldsymbol{u}_{1+1/2}$. Again, by using $\boldsymbol{u}_1$, $\dot{\boldsymbol{u}}_1$ and $\boldsymbol{u}_{1+1/2}$ we obtain $\boldsymbol{u}_2$ by using the second vademecum. This procedure is repeated for each time step of the simulation. This loop indeed runs under very astringent real time constraints, such as those of a realistic rendering.

As in previous examples, we assume for simplicity of the exposition, that the load is of unity module and acts along the vertical axis: $\boldsymbol{t} = \boldsymbol{e}_k \cdot \delta(\boldsymbol{x} - \boldsymbol{s})$, where δ represents the Dirac-delta function and $\boldsymbol{e}_k$ the unit vector along the z-coordinate axis on the top of the domain. A more general setting would need for new parameters, i.e., the components of the load vector, for instance, but it is perfectly possible in the same framework here explained.

Regarding the matrix structure of the problems given by Eqs. (5.5) and (5.6), in both of them the force vector applied at each time step, $\boldsymbol{F}_{n+1/2}$ and $\boldsymbol{F}_{n+1}$ appears. Like in the parametric problem in Chap. 3, we must consider the load in a separated form, i.e. a separated sum of products of separated functions,

$$f_{n+1/2}(\boldsymbol{x}, \boldsymbol{\zeta}_n, \dot{\boldsymbol{\zeta}}_n, \ddot{\boldsymbol{\zeta}}_n, p_a, \boldsymbol{s}) = \sum_{j=1}^{m} \boldsymbol{f}_x^j \cdot \boldsymbol{f}_{\zeta_n}^j \cdot \boldsymbol{f}_{\dot{\zeta}_n}^j \cdot \boldsymbol{f}_{\ddot{\zeta}_n}^j \cdot \boldsymbol{f}_{p_c}^j \cdot \boldsymbol{f}_s^j. \tag{5.10}$$

A simple way to obtain such a decomposition is to consider as many terms j as possible nodal load locations, and to set $\boldsymbol{f}_x^j$, as the force modulus (here, unity). In turn, $\boldsymbol{f}_{p_c}^j = [0\ 1]^T$ for each j, $\boldsymbol{f}_s^j = \boldsymbol{I}$ (the identity matrix) and the rest of vectors $\boldsymbol{f}_{\zeta_n}^j$, $\boldsymbol{f}_{\dot{\zeta}_n}^j$, $\boldsymbol{f}_{\ddot{\zeta}_n}^j = \mathbf{1}$, that is, the `ones` instruction in Matlab, a vector composed by ones in every entry. We proceed analogously for the force vector $\boldsymbol{f}_{n+1}$ in the second substep of the time integrator scheme.

In order to completely define the right-hand side vector, let us see how to compute the others terms in Eqs. (5.5) and (5.6). In these formulae, we find terms defined as a constant value multiplying the mass matrix $\boldsymbol{M_m}$ and multiplying $\boldsymbol{\zeta}_n$, $\dot{\boldsymbol{\zeta}}_n$, $\ddot{\boldsymbol{\zeta}}_n$ and $\boldsymbol{u}_{n+1}$. These vectorial parameters should be considered in separated form, so as to have the following form,

$$\begin{aligned}
\zeta_n &= \bar{\mathbf{1}}_x \cdot \left[\zeta_n^{min} \ldots \zeta_n^{max}\right]^T \cdot \mathbf{1}_{\dot{\zeta}_n} \cdot \mathbf{1}_{\ddot{\zeta}_n} \cdot \mathbf{1}_{p_c} \cdot \mathbf{1}_s, \\
\dot{\zeta}_n &= \bar{\mathbf{1}}_x \cdot \mathbf{1}_{\zeta_n} \cdot \left[\dot{\zeta}_n^{min} \ldots \dot{\zeta}_n^{max}\right]^T \cdot \mathbf{1}_{\ddot{\zeta}_n} \cdot \mathbf{1}_{p_c} \cdot \mathbf{1}_s, \\
\ddot{\zeta}_n &= \bar{\mathbf{1}}_x \cdot \mathbf{1}_{\zeta_n} \cdot \mathbf{1}_{\dot{\zeta}_n} \cdot \left[\ddot{\zeta}_n^{min} \ldots \ddot{\zeta}_n^{max}\right]^T \cdot \mathbf{1}_{p_c} \cdot \mathbf{1}_s, \\
\zeta_{n+1/2} &= \bar{\mathbf{1}}_x \cdot \mathbf{1}_{\zeta_n} \cdot \mathbf{1}_{\dot{\zeta}_n} \cdot \left[\zeta_{n+1/2}^{min} \ldots \zeta_{n+1/2}^{max}\right]^T \cdot \mathbf{1}_{p_c} \cdot \mathbf{1}_s,
\end{aligned}$$

where vector $\bar{\mathbf{1}}_x$ refers to the ones vector in space direction that satisfies essential boundary conditions or, equivalently, the ones vector in which entries related to nodes pertaining to the essential boundary have been replaced by zeros.

In the code reproduced below, these values of the RHS vector are computed for each reduced basis `nrb` and at each substep, and are saved in the $\boldsymbol{FR1}$ and $\boldsymbol{FR2}$ vectors. So for substep 1,

$$\begin{aligned}
\boldsymbol{FR1}_1^j &= \bar{\mathbf{1}}_x \cdot \left[\zeta_n^{min} \ldots \zeta_n^{max}\right]^T \cdot \mathbf{1}_{\dot{\zeta}_n} \cdot \mathbf{1}_{\ddot{\zeta}_n} \cdot \mathbf{1}_{p_c} \cdot \mathbf{1}_s, \\
\boldsymbol{FR1}_2^j &= \bar{\mathbf{1}}_x \cdot \mathbf{1}_{\zeta_n} \cdot \left[\dot{\zeta}_n^{min} \ldots \dot{\zeta}_n^{max}\right]^T \cdot \mathbf{1}_{\ddot{\zeta}_n} \cdot \mathbf{1}_{p_c} \cdot \mathbf{1}_s, \\
\boldsymbol{FR1}_3^j &= \bar{\mathbf{1}}_x \cdot \mathbf{1}_{\zeta_n} \cdot \mathbf{1}_{\dot{\zeta}_n} \cdot \left[\ddot{\zeta}_n^{min} \ldots \ddot{\zeta}_n^{max}\right]^T \cdot \mathbf{1}_{p_c} \cdot \mathbf{1}_s,
\end{aligned}$$

where $j = 1, \ldots,$ `nrb` refers to the number in the set of reduced basis.

Equivalently, in substep 2,

$$\begin{aligned}
\boldsymbol{FR2}_1^j &= \bar{\mathbf{1}}_x \cdot \left[\zeta_n^{min} \ldots \zeta_n^{max}\right]^T \cdot \mathbf{1}_{\dot{\zeta}_n} \cdot \mathbf{1}_{\ddot{\zeta}_n} \cdot \mathbf{1}_{p_c} \cdot \mathbf{1}_s, \\
\boldsymbol{FR2}_2^j &= \bar{\mathbf{1}}_x \cdot \mathbf{1}_{\zeta_n} \cdot \left[\dot{\zeta}_n^{min} \ldots \dot{\zeta}_n^{max}\right]^T \cdot \mathbf{1}_{\ddot{\zeta}_n} \cdot \mathbf{1}_{p_c} \cdot \mathbf{1}_s, \\
\boldsymbol{FR2}_3^j &= \bar{\mathbf{1}}_x \cdot \mathbf{1}_{\zeta_n} \cdot \mathbf{1}_{\dot{\zeta}_n} \cdot \left[\zeta_n^{min} \ldots \zeta_n^{max}\right]^T \cdot \mathbf{1}_{p_c} \cdot \mathbf{1}_s,
\end{aligned}$$

where j refers to the number of reduced basis. We use the same discretization rationale for $\boldsymbol{u}_n$ and $\boldsymbol{u}_{n+1/2}$.

To solve both substeps so as to generate the multi-parametric solution, we employ a greedy algorithm, using a fixed-point strategy, so as to compute the new terms in the sum, represented by Eqs. (5.8) and (5.9). If, within the enrichment loop, the solution is not accurate enough, the already computed approximation is improved by adding a new separated term

$$\boldsymbol{u}_{n+1/2} = \sum_{i=1}^{N} \prod_{j=1}^{3+3*\mathtt{nrb}} \boldsymbol{N}^T(\text{var j})\boldsymbol{F1}^i_{\text{var j}} + \prod_{j=1}^{3+3*\mathtt{nrb}} \boldsymbol{N}^T(\text{var j})\boldsymbol{R}^i_{\text{var j}}, \tag{5.11}$$

where N is the number of terms already for the PGD solution. This Eq. (5.11) is analogous for substep 2, with just changing $\boldsymbol{u}_{n+1/2}$ by $\boldsymbol{u}_{n+1}$, and $\boldsymbol{F1}^i_{\text{var j}}$ by $\boldsymbol{F2}^i_{\text{var j}}$.

Finally, Eq. (5.5) is solved in the code written below using the following notation:

$$\begin{aligned}
&\prod_{j=1}^{3+3*\mathtt{nrb}} \boldsymbol{R}^*_j \left[\left[\frac{16}{\Delta t^2}\right]\boldsymbol{M_m} + \boldsymbol{K_m}\right] \prod_{j=1}^{3+3*\mathtt{nrb}} \boldsymbol{R}_j \\
&= \prod_{j=1}^{3+3*\mathtt{nrb}} \boldsymbol{R}^*_j \cdot \left[\boldsymbol{F}_{n+1/2} + \left[\frac{16}{\Delta t^2}\right]\boldsymbol{M_m} \cdot \boldsymbol{FR1}_1 + \left[\frac{8}{\Delta t}\right]\boldsymbol{M_m} \cdot \boldsymbol{FR1}_2 + \boldsymbol{M_m} \cdot \boldsymbol{FR1}_3\right. \\
&\qquad \left. - \sum_{i=1}^{N}\left[\left[\frac{16}{\Delta t^2}\right]\boldsymbol{M_m} + \boldsymbol{K_m}\right] \prod_{j=1}^{3+3*\mathtt{nrb}} \boldsymbol{F1}^i_j\right]
\end{aligned} \tag{5.12}$$

where $\prod_{j=1}^{3+3*\mathtt{nrb}} \boldsymbol{R}^*_j$ represents the weight function, that within the fixed-point strategy takes a different form depending on the particular iteration. Thus if, for instance, we are computing along the k-th coordinate, assuming the others directions to be known, we have

$$\prod_{j=1}^{3+3*\mathtt{nrb}} \boldsymbol{R}^*_j = \prod_{j=1, j\neq k}^{3+3*\mathtt{nrb}} \boldsymbol{R}_j \cdot \boldsymbol{R}^*_k. \tag{5.13}$$

These variables can be readily identified in the routine `enrichment_substep1` by taking into account the following notation,

$$\begin{aligned}
&\mathtt{matrix1} = \left[\frac{16}{\Delta t^2}\right]\boldsymbol{M_m}, \\
&\mathtt{matrix2} = \boldsymbol{K_m}, \\
&\boldsymbol{V} = \boldsymbol{F}_{n+1/2}, \\
&\mathtt{value1} = \left[\frac{16}{\Delta t^2}\right]\boldsymbol{M_m} \cdot \boldsymbol{FR1}_1, \\
&\mathtt{value2} = \left[\frac{8}{\Delta t}\right]\boldsymbol{M_m} \cdot \boldsymbol{FR1}_2, \\
&\mathtt{value3} = \boldsymbol{M_m} \cdot \boldsymbol{FR1}_3, \\
&\boldsymbol{FV} = \boldsymbol{F1}.
\end{aligned}$$

For Eq. (5.6), the implemented routine `enrichment_substep2` computes

$$\prod_{j=1}^{3+3*\mathtt{nrb}} \boldsymbol{R}_j^* \left[\left[\frac{9}{\Delta t^2} \right] \boldsymbol{M}_m + \boldsymbol{K}_m \right] \prod_{j=1}^{3+3*\mathtt{nrb}} \boldsymbol{R}_j$$

$$= \prod_{j=1}^{3+3*\mathtt{nrb}} \boldsymbol{R}_j^* \cdot \left[\boldsymbol{F}_{n+1} - \left[\frac{19}{\Delta t^2} \right] \boldsymbol{M}_m \cdot \boldsymbol{FR2}_1 - \left[\frac{5}{\Delta t} \right] \boldsymbol{M}_m \cdot \boldsymbol{FR2}_2 + \left[\frac{28}{\Delta t^2} \right] \boldsymbol{M}_m \cdot \boldsymbol{FR2}_3 \right.$$

$$\left. - \sum_{i=1}^{N} \left[\left[\frac{16}{\Delta t^2} \right] \boldsymbol{M}_m + \boldsymbol{K}_m \right] \prod_{j=1}^{3+3*\mathtt{nrb}} \boldsymbol{F1}_j^i \right].$$

In turn, $\prod_{j=1}^{3+3*nrb} \boldsymbol{R}_j^*$ is defined in Eq. (5.13). The variables can be identified as,

$$\mathtt{matrix1} = \left[\frac{9}{\Delta t^2} \right] \boldsymbol{M}_m,$$

$$\mathtt{matrix2} = \boldsymbol{K}_m,$$

$$\boldsymbol{V} = \boldsymbol{F}_{n+1},$$

$$\mathtt{value1} = \left[\frac{19}{\Delta t^2} \right] \boldsymbol{M}_m \cdot \boldsymbol{FR2}_1,$$

$$\mathtt{value2} = \left[\frac{5}{\Delta t} \right] \boldsymbol{M}_m \cdot \boldsymbol{FR2}_2,$$

$$\mathtt{value3} = \left[\frac{28}{\Delta t^2} \right] \boldsymbol{M}_m \cdot \boldsymbol{FR2}_3,$$

$$\boldsymbol{FV} = \boldsymbol{F2}.$$

In next section the detailed code implmenting this strategy is provided.

5.4 Matlab Code

As always, the code begins by the `main.m` file, which is reproduced below. In this case, a series of previous simulations are needed so as to construct the POD basis referred to in Eq. (5.7). These simulations were carried out by us with the help of the commercial software Abaqus, although the reader can use his or her preferred code to do it. Once these simulations are done, and the POD modes computed, they are stored in memory by means of the instruction `WS = load('WorkSpaceBeam_REF.mat','Vreal');`, see below.

The quality of the final results will obviously depend on the similarity of this POD basis to the problem being simulated. In general, our experience indicates that with good basis, the number of POD modes necessary for a good energy conservation (i.e., avoidance of numerical dissipation) tends to be on the order of 6–8 modes.

```
%
%                          PGD Code for Dynamic Problem
%                        D. González, I. Alfaro, E. Cueto
%                            Universidad de Zaragoza
%                              AMB-I3A Dec 2015
%
clear all; close all; clc;
%
% VARIABLES
%
global E nu coords tet
Modulus = 10000.0; % Force modulus.
cooru = linspace(-5E1,5E1,300); % Discretization for displacement field.
coorv = linspace(-5E+2,5E+2,300);% Discretization for velocity field.
coora = linspace(-1E+3,1E+3,300); % Discretization for acceleration field.
deltat = 0.00125; coort = 0:deltat:2; % Time discretization.
TOL = 1.0E-04; % Tolerance.
num_max_iter = 15; % # of summands of the approach.
E = 2E11; nu = 0.3; Rho = 2.5E+04; % Material.
deltatA = 0.00125; % Time step for Reference problem - Abaqus' result.
NodeR = 6; NodeC = 104; % Reference nodes to compare PGD solution.
nrb = 1; % Number of directions on reduced basis.
% The PGD solution depends on: Space, Load, load parameter and
% displacement-velocity-acceleration for each reduced POD basis
nv = 3 + 3*nrb; % # of parameters (or variables) for the PGD solution.
%
% GEOMETRY
%
coords = load('gcoordBeam.dat'); % Nodal coordinates.
tet = load('conecBeam.dat'); % Connectivity list.
Ind = 1:size(coords,1); % List of nodes.
bcnode = Ind(coords(:,1)==min(coords(:,1))); % Boundary: Fix left side.
IndBcnode = sort([3*(bcnode-1)+1 3*(bcnode-1)+2 3*bcnode]); % D.o.f. BCs.
dof = setxor(IndBcnode,1:numel(coords))'; % D.o.f. Free nodes.
% Make use of triangulation MatLab function to obtain boundary surface.
TR = triangulation(tet,coords);
[tf] = freeBoundary(TR); % Dependent of 3D geometry of the boundary.
[tri,coors] = freeBoundary(TR); % Independent triangulation of boundary.
IndS = 1:size(coors,1); ncoors = numel(IndS);
%
% STIFNESS AND MASS MATRICES COMPUTATION
%
[r1,r2] = fem3D; r2 = Rho.*r2; % Space: Stiffness: r1, Mass: r2.
[z1,vu] = elemstiff(cooru); % Displacements.
[w1,vv] = elemstiff(coorv); % Velocities.
[s1,va] = elemstiff(coora); % Accelerations.
vu = repmat(vu,1,ncoors); % Reshape to construct source in separated form.
vv = repmat(vv,1,ncoors);
va = repmat(va,1,ncoors);
%
% SOURCE
%
coorp = 1:size(coors,1); % We consider each possible load position
                                             % as a different load case.
p1 = elemstiff(coorp); % Mass matrix for load parameter: 1st vargout.
% Identifying local nodes of the loaded surface on the global connectivity.
% To obtain that: coords(IndL,:)-coors = zeros(nn2,1).
[trash,trash2,xj] = intersect(IndS,tri(:)); % TRI Local connectivity.
IndL = tf(xj); % TF Global connectivity of the loaded surface (Top).
DOFLoaded = 3*IndL; % Consider vertical load on the top of the beam.
vx = zeros(numel(coords),ncoors);
vx(DOFLoaded,:) = eye(ncoors); % Space terms for the source.
vp = eye(ncoors); vp = p1*vp; % Load terms for the source.
%
% ACTIVATION OF LOAD PARAMETER
%
```

```
coorac = [1 2]; % Value to activate the load. Two possibilities 1-No 2-Yes.
pal = elemstiff(coorac); % Mass matrix for load activation: 1st vargout.
vpa = pal*repmat([0; -Modulus],1,ncoors); % Activation terms for the source
%
% LOADING P.O.D. DATA TO CONSTRUCT REDUCED BASIS
%
WS = load('WorkSpaceBeam_REF.mat','Vreal');
Vreal = WS.Vreal; % Loading Displacement field of the reference solution.
%
% APPLY P.O.D. TECHNIQUE TO OBTAIN REDUCED BASIS
%
Q = Vreal(dof,:)*Vreal(dof,:)'; [A,lam] = eigs(Q,[],nrb);
%
% ALLOCATION OF MATRICES AND VECTORS FOR EACH TIME INTEGRATION STEP (1,2)
%
K1 = cell(nv,1); M1 = K1; V1 = K1; Fv1 = K1; FR1 = K1; coor1 = cell(nv,1);
K2 = cell(nv,1); M2 = K2; V2 = K2; Fv2 = K2; FR2 = K2; coor2 = cell(nv,1);
%
% SPACE MATRICES
%
K1{1} = r1; % Stiffness matrix for SubStep 1.
M1{1} = r2; % Mass matrix for SubStep 1.
V1{1} = vx; % Space term for the source at the SubStep 1.
K2{1} = r1; M2{1} = r2; V2{1} = vx; % SubStep 2.
%
% LOAD MATRICES
%
K1{nv} = p1; % Stiffness matrix contribution of load for SubStep 1.
M1{nv} = p1; % Mass matrix for SubStep 1.
V1{nv} = vp; % Load term for the source at the SubStep 1.
coor1{nv} = coorp; % Load Discretization for SubStep 1.
K2{nv} = p1; M2{nv} = p1; V2{nv} = vp; coor2{nv} = coorp; % SubStep 2.
%
% MATRICES RELATED TO ACTIVATION PARAMETER
%
K1{nv-1} = pal; % Stiffness contribution of act.param. for SubStep 1.
M1{nv-1} = pal; % Mass matrix for SubStep 1.
V1{nv-1} = vpa; % Activation parameter term for the source, SubStep 1.
coor1{nv-1} = coorac; % Activation parameter discretization for SubStep 1.
K2{nv-1} = pal; M2{nv-1} = pal; V2{nv-1} = vpa; coor2{nv-1} = coorac; % S2.
%
% REDUCED BASIS MATRICES
%
for i1=2:3:nv-2
    %
    % SUBSTEP 1
    %
    K1{i1} = z1; M1{i1} = z1; V1{i1} = vu; coor1{i1} = cooru; % U
    K1{i1+1} = w1; M1{i1+1} = w1; V1{i1+1} = vv; coor1{i1+1} = coorv; % V
    K1{i1+2} = s1; M1{i1+2} = s1; V1{i1+2} = va; coor1{i1+2} = coora; % A
    %
    % SUBSTEP 2
    %
    K2{i1} = z1; M2{i1} = z1; V2{i1} = vu; coor2{i1} = cooru; % U
    K2{i1+1} = w1; M2{i1+1} = w1; V2{i1+1} = vv; coor2{i1+1} = coorv; % V
    K2{i1+2} = z1; M2{i1+2} = z1; V2{i1+2} = vu; coor2{i1+2} = cooru; % U/2
end
%
% INICIALIZATING PGD SOLUTION
%
for i1=1:nv
    Fv1{i1} = 0.0.*V1{i1}(:,1); % PGD vectors for SubStep 1.
    Fv2{i1} = 0.0.*V2{i1}(:,1); % PGD vectors for SubStep 2.
end
%
% BOUNDARY CONDITIONS
%
```

```
Free1 = cell(nv,1); Free2 = cell(nv,1);
Free1{1} = dof; Free2{1} = dof; % Free DOF for Space.
for i1=2:nv % No BCs for rest of parameters (variables).
    Free1{i1} = 1:numel(coor1{i1});
    Free2{i1} = 1:numel(coor2{i1});
end
%
% Un, Vn, An ... IN SEPARATED FORM FOR INTEGRATION SCHEME
%
% We have 3*nen terms in the source related to Un, Vn and An in SubStep 1
% and Vn, Un+1/2 and Un for SubStep 2. Following the sort of the variables
% in the PGD solution for SubStep 1, U1_{n+1/2}(Var_1, Var_2,...,
% Var_{nv-1}, Var_{nv}), where Var_1=Spatial Coordinates, Var_{nv-1} =
% Activation Parameter, Var_{nv} = Loads, and Var_(2:3:nv_1) = U_n
% (Displacement in time n), Var_(3:3:nv_1) = V_n (Velocity in time n) and
% Var_(4:3:nv_1) = A_n (Acceleration in time n)
% For the PGD solution for SubStep 2, U2_{n+1/2}(Var_1, Var_2,...,
% Var_{nv-1}, Var_{nv}), where Var_1=Spatial Coordinates, Var_{nv-1} =
% Activation Parameter, Var_{nv} = Loads, and Var_(2:3:nv_1) = U_n
% (Displacement in time n), Var_(3:3:nv_1) = V_n (Velocity in time n) and
% Var_(4:3:nv_1) = U_{n+1/2} (Displacement in time n + 1/2).
%
% IMPORTANT: To obtain U_n and V_n variables in SubStep 1 (for instance) in
% separated form we consider that:
% U_n = 1_{space} U_n 1_{velocity} 1_{acceleration} 1_{activation} 1_{load}
% V_n = 1_{space} 1_{displac.} V_n 1_{acceleration} 1_{activation} 1_{load}
FR1{1} = zeros(size(Fv1{1},1),nv-3); FR2{1} = zeros(size(Fv2{1},1),nv-3);
for i1=2:nv
    FR1{i1} = ones(size(Fv1{i1},1),nv-3);
    FR2{i1} = ones(size(Fv2{i1},1),nv-3);
end
for i1=1:3 % 3 terms per # reduced basis.
    FR1{1}(dof,i1:3:end) = A; % Projection space onto Reduced basis.
    FR2{1}(dof,i1:3:end) = A;
end
for i1=2:nv-2
    FR1{i1}(:,i1-1) = coor1{i1}'; % U_n, V_n, A_n.
    FR2{i1}(:,i1-1) = coor2{i1}'; % U_n, V_n, U_{n+1/2}.
end
%
% ENRICHMENT OF THE APPROXIMATION: SUBSTEP 1
%
num_iter1 = 0; Error_iter = 1.0; iter = zeros(1); Aprt = 0;
while Error_iter>TOL && num_iter1<num_max_iter
    num_iter1 = num_iter1 + 1; R0 = cell(nv,1);
    for i1=1:nv
        R0{i1} = ones(size(Fv1{i1},1),1); % Initial guess for R, S, ...
    end;
    R0{1}(IndBcnode) = 0; % We impose initial guess for spacial coordinates
    %
    % ENRICHMENT STEP
    %
    [R,iter(num_iter1)] = enrichment_substep1(K1,M1,V1,num_iter1,Fv1,R0,...
        FR1,Free1,deltat);
    for i1=1:nv, Fv1{i1}(:,num_iter1) = R{i1}; end % R is valid, add it.
    %
    % STOPPING CRITERION
    %
    Error_iter = 1.0;
    for i1=1:nv
        Error_iter = Error_iter.*norm(Fv1{i1}(:,num_iter1));
    end
    Aprt = max(Aprt,sqrt(Error_iter));
    if num_iter1>nrb, Error_iter = sqrt(Error_iter)/Aprt; end
    fprintf(1,'SubStep␣1:␣%dst␣summand␣in␣%d␣',num_iter1,iter(num_iter1));
    fprintf(1,'iterations␣with␣a␣weight␣of␣%f\n',sqrt(Error_iter));
end
```

```
fprintf(1,'\n');
%
% ENRICHMENT OF THE APPROXIMATION: SUBSTEP 2
%
num_iter2 = 0; Error_iter = 1.0; iter = zeros(1); Aprt = 0;
while Error_iter>TOL && num_iter2<num_max_iter
    num_iter2 = num_iter2 + 1; R0 = cell(nv,1);
    for i1=1:nv
        R0{i1} = rand(size(Fv2{i1},1),1); % Initial guess for R, S, ...
    end
    R0{1}(IndBcnode) = 0; % We impose initial guess for spacial coordinates
    %
    % ENRICHMENT STEP
    %
    [R,iter(num_iter2)] = enrichment_substep2(K2,M2,V2,num_iter2,Fv2,R0,...
        FR2,Free2,deltat);
    for i1=1:nv, Fv2{i1}(:,num_iter2) = R{i1}; end % R is valid, add it.
    %
    % STOPPING CRITERION
    %
    Error_iter = 1.0;
    for i1=1:nv
        Error_iter = Error_iter.*norm(Fv2{i1}(:,num_iter2));
    end
    Aprt = max(Aprt,sqrt(Error_iter));
    if num_iter2>nrb, Error_iter = sqrt(Error_iter)/Aprt; end
    fprintf(1,'SubStep␣2:␣%dst␣summand␣in␣%d␣',num_iter2,iter(num_iter2));
    fprintf(1,'iterations␣with␣a␣weight␣of␣%f\n',sqrt(Error_iter));
end
fprintf(1,'PGD␣Process␣exited␣normally\n\n');
save('WorkSpacePGD_Dynamic.mat');
%
% POST-PROCESSING
%
figure; % Plotting reference solution for the node NodeR
plot(0:deltatA:deltatA*(size(Vreal,2)-1),Vreal(3*(NodeR-1)+3,:),...
    'b-','LineWidth',2.5); % Reference solution for vertical displacement
%
% ALLOCATE MEMORY FOR SUBSTEP SOLUTION VECTORS
%
Mv = cell(nv,1); % Nodal values for each parameter
% We need to compute SUBSTEP 1: MůA_{n+1/2} + KůU_{n+1/2} = F_{n+1/2}
% and A_{n+1/2} = 4*(V_{n+1/2}-V_n)/Deltat - A_n
% and V_{n+1/2} = 4*(U_{n+1/2}-U_n)/Deltat - V_n
Disp2 = zeros(nrb,numel(coort)); % Allocate memory for U_{n+1/2}
Vel2 = zeros(nrb,numel(coort)); % Allocate memory for V_{n+1/2}
Acel2 = zeros(nrb,numel(coort)); % Allocate memory for A_{n+1/2}
% We need to compute SUBSTEP 2: MůA_{n+1} + KůU_{n+1} = F_{n+1}
% and A_{n+1} = V_n/Deltat - 4*V_{n+1/2}/Deltat + 3*V_{n+1}/Deltat
% and V_{n+1} = U_n/Deltat - 4*U_{n+1/2}/Deltat + 3*U_{n+1}/Deltat
Disp = zeros(nrb,numel(coort)); % Allocate memory for U_{n+1}
Vel = zeros(nrb,numel(coort)); % Allocate memory for V_{n+1}
Acel = zeros(nrb,numel(coort));% Allocate memory for A_{n+1}
% We compute U_n,V_n,... onto reduced basis. We back to real Space
RealDisp2 = zeros(numel(coords),numel(coort));
RealDisp = zeros(numel(coords),numel(coort));
%
% INITIAL VALUES FOR DISPLACEMENT FIELD
%
Ai = pinv(A); % Pseudo-inverse P.O.D. matrix.
% 1st and 2nd displacement field for t = 0 and t = deltat from reference
% solution to start the simulation.
RealDisp(dof,1:2) = Vreal(dof,1:2).*deltat./deltatA;
% We project these first two displacements onto reduced basis.
Disp(:,1:2) = Ai*RealDisp(dof,1:2).*deltat./deltatA;
% We interpolate the reference solution for [1:2]+1/2 steps.
Disp2(:,1) = Ai*((Vreal(dof,1) + Vreal(dof,2))/2).*deltat./deltatA;
```

```
Disp2(:,2) = Ai*((Vreal(dof,2) + Vreal(dof,3))/2).*deltat./deltatA;
%
% REAL-TIME LOOP FOR TIME INTEGRATION
%
for k2=3:numel(coort)
    %
    % APPLY SUBSTEP 1
    %
    for i1=2:3:nv-2
        Mv{i1} = evaluate_shpfunc(coor1{i1},Disp((i1+1)/3,k2-1));
        Mv{i1+1} = evaluate_shpfunc(coor1{i1+1},Vel((i1+1)/3,k2-1));
        Mv{i1+2} = evaluate_shpfunc(coor1{i1+2},Acel((i1+1)/3,k2-1));
    end
    Mv{nv} = evaluate_shpfunc(coor1{nv},NodeC); % Load case NodeC.
    if coort(k2)-deltat/2<0.25
        Mv{nv-1} = [0; 1]; % Value of the shape function for act. parameter
    elseif coort(k2)-deltat/2>=0.5 % Time when the load vanishes
        Mv{nv-1} = [1; 0]; % Value of the shape function for act. parameter
    else % Ramp case
        Mv{nv-1}(2) = (0.5-(coort(k2)-deltat/2))/0.25;
        Mv{nv-1}(1) = 1-Mv{nv-1}(2); % Value of the shape function.
    end
    %
    % COMPUTE PGD SUBSTEP 1 SOLUTION
    %
    for k1=1:num_iter1
        value1 = Fv1{1}(dof,k1);
        for j1=2:nv, value1 = value1.*(Mv{j1}'*Fv1{j1}(:,k1)); end
        RealDisp2(dof,k2) = RealDisp2(dof,k2) + value1;
    end
    %
    % TIME SCHEME SUBSTEP 1
    %
    % MůA_{n+1/2} + KůU_{n+1/2} = F_{n+1/2}
    % and A_{n+1/2} = 4*(V_{n+1/2}-V_n)/Deltat - A_n
    % and V_{n+1/2} = 4*(U_{n+1/2}-U_n)/Deltat - V_n
    Disp2(:,k2) = Ai*RealDisp2(dof,k2); % Project onto reduced basis
    Vel2(:,k2) = 4*(Disp2(:,k2)-Disp(:,k2-1))/deltat - Vel(:,k2-1);
    Acel2(:,k2) = 4*(Vel2(:,k2)-Vel(:,k2-1))/deltat - Acel(:,k2-1);
    %
    % APPLY SUBSTEP 2
    %
    for i1=2:3:nv-2
        Mv{i1+2} = evaluate_shpfunc(coor2{i1+2},Disp2((i1+1)/3,k2));
    end
    Mv{nv} = evaluate_shpfunc(coor1{nv},NodeC); % Load case NodeC.
    if coort(k2)<0.25
        Mv{nv-1} = [0; 1]; % Value of the shape function for act. parameter
    elseif coort(k2)>=0.5 % Time when the load vanishes
        Mv{nv-1} = [1; 0]; % Value of the shape function for act. parameter
    else % Ramp case
        Mv{nv-1}(2) = (0.5-coort(k2))/0.25;
        Mv{nv-1}(1) = 1-Mv{nv-1}(2); % Value of the shape function.
    end
    %
    % COMPUTE PGD SUBSTEP 2 SOLUTION
    %
    for k1=1:num_iter2
        value2 = Fv2{1}(dof,k1);
        for j1=2:nv, value2 = value2.*(Mv{j1}'*Fv2{j1}(:,k1)); end
        RealDisp(dof,k2) = RealDisp(dof,k2) + value2;
    end
    %
    % TIME SCHEME SUBSTEP 2
    %
    % MůA_{n+1} + KůU_{n+1} = F_{n+1}
    % and A_{n+1} = V_n/Deltat - 4*V_{n+1/2}/Deltat + 3*V_{n+1}/Deltat
```

```
    % and V_{n+1} = U_n/Deltat - 4*U_{n+1/2}/Deltat + 3*U_{n+1}/Deltat
    Disp(:,k2) = Ai*RealDisp(dof,k2); % Project onto reduced basis
    Vel(:,k2) = Disp(:,k2-1)/deltat - 4*Disp2(:,k2)/deltat + ...
        3*Disp(:,k2)/deltat;
    Acel(:,k2) = Vel(:,k2-1)/deltat - 4*Vel2(:,k2)/deltat + ...
        3*Vel(:,k2)/deltat;
end
%
% PLOT PGD FINAL SOLUTION AND COMPARE WITH THE REFERENCE SOLUTION
%
hold on; plot(coort,RealDisp(3*(NodeR-1)+3,:),'m--','LineWidth',2.5);
save('WorkSpacePGD_Dynamic.mat');
fprintf(1,'\n#####ăEnd␣of␣simulation␣#####\n\n');
```

This just presented code reads nodal coordinates and connectivity list from files `gcoordBeam.dat` and `conecBeam.dat`, respectively. If, after the execution of the program, one types

» `trisurf(tri,coors(:,1),coors(:,2),coors(:,3));`

» `axis equal`

in Matlab's command line, a plot of the finite element model of the beam is obtained, see Fig. 5.2.

As in previous examples, the code calls to the subroutine `elemstiff.m`, which is reproduced below.

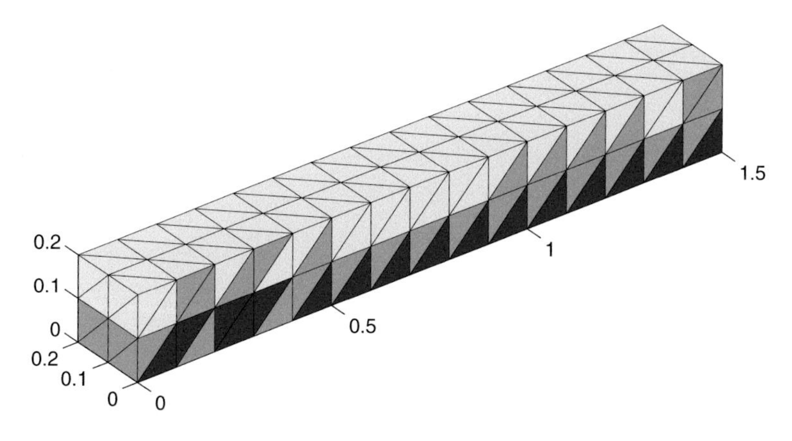

Fig. 5.2 Finite element mesh for the beam dynamics problem. The beam is assumed to be encastred at $y = 0$

```
function [p1,p2] = elemstiff(coor)
% function [p1,p2] = elemstiff(coor)
% For variable X compute p1=\int N N, p2=\int N
% Universidad de Zaragoza - 2015
nen = numel(coor);
%
% ALLOCATE MEMORY
%
p1 = zeros(nen); p2 = zeros(nen,1);
X = coor(1:nen-1)'; Y = coor(2:nen)'; % Coordinates of elements.
L = Y - X; % Longitude of each  element for parametrized variable.
sg = [-1.0/sqrt(3.0) 1.0/sqrt(3.0)]; wg = ones(2,1); % Gauss points.
npg = numel(sg);
for i1=1:nen-1
    c = zeros(1,npg); c(1,:) = 0.5.*(1.0-sg).*X(i1) + 0.5.*(1.0+sg).*Y(i1);
    N = zeros(nen,npg);
    N(i1+1,:) = (c(1,:)-X(i1))./L(i1); N(i1,:) = (Y(i1)-c(1,:))./L(i1);
    for j1=1:npg
        p1 = p1 + N(:,j1)*N(:,j1)'*0.5.*wg(j1).*L(i1); % \int N N.
        p2 = p2 + N(:,j1).*0.5.*wg(j1).*L(i1); % \int N.
    end
end
return
```

Our particular implementation of the method makes use of the energy and momentum conserving algorithm by Bathe [15]. This algorithm makes use of a predictor-corrector algorithm, whose first sub-step is included in routine `enrichment_substep1`, reproduced here:

```
function [R,iter] = enrichment_substep1(K,M,V,num_iter,FV,R,FR,CC,deltat)
% [R,iter] = enrichment_substep1(K,M,V,num_iter,FV,R,FR,CC,deltat)
% Compute the new R, S, functions to enrich the PGD solution for SubStep 1
% in Dynamic problems.
% Universidad de Zaragoza - 2015
iter = 1; TOL = 1.0E-4; error = 1.0; % Initializating values.
nv = size(FV,1); % # Number of variables for the PGD.
mxit = 11; % # of possible iterations for the fixed point algorithm.
%
% FIXED POINT ALGORITHM
%
while abs(error)>TOL
    Raux = R; % Updating R(S) last values.
    for i1=1:nv
        %
        % MATRIX COMPUTATION
        %
        matrix1 = 16.0/deltat/deltat; % Constant of mass contribution.
        matrix2 = 1.0; % Constant of stiffness contribution.
        for j1=1:nv
            if j1~=i1
                matrix1 = matrix1.*(R{j1}'*M{j1}*R{j1});
                matrix2 = matrix2.*(R{j1}'*K{j1}*R{j1});
            end
        end
        matrix = M{i1}.*matrix1 + K{i1}.*matrix2;
        %
        % SOURCE COMPUTATION
        %
        source = 0.0;
        for k1=1:size(V{1},2) % Loop over number of functions of the source
            sourceval = 1.0;
            for j1=1:nv
                if j1~=i1
                    sourceval = sourceval.*(R{j1}'*V{j1}(:,k1));
```

```
                end
            end
            source = source + sourceval.*V{i1}(:,k1);
        end
        for k1=1:3:(nv-3)
            value1 = 16/deltat/deltat;% Constant value for U_n contribution
            value2 = 8/deltat; % Constant value for V_n contribution.
            value3 = 1.0; % Constant value for A_n contribution.
            for j1=1:nv
                if j1~=i1
                    value1 = value1.*(R{j1}'*M{j1}*FR{j1}(:,k1));
                    value2 = value2.*(R{j1}'*M{j1}*FR{j1}(:,k1+1));
                    value3 = value3.*(R{j1}'*M{j1}*FR{j1}(:,k1+2));
                end
            end
            source = source + value1.*(M{i1}*FR{i1}(:,k1)) + ...
                value2.*(M{i1}*FR{i1}(:,k1+1)) + ...
                value3.*(M{i1}*FR{i1}(:,k1+2));
        end
        %
        % CONTRIBUTION TO SOURCE OF KNOWNING SOLUTION
        %
        for i2=1:num_iter-1
            value1 = 16/deltat/deltat; % Constant of mass contribution.
            value2 = 1.0; % Constant of stiffness contribution.
            for j1=1:nv
                if j1~=i1
                    value1 = value1.*(R{j1}'*M{j1}*FV{j1}(:,i2));
                    value2 = value2.*(R{j1}'*K{j1}*FV{j1}(:,i2));
                end
            end
            source = source - (M{i1}*FV{i1}(:,i2)).*value1 - ...
                (K{i1}*FV{i1}(:,i2)).*value2;
        end
        %
        % SOLVE THE R{i1} VARIABLE
        %
        R{i1}(CC{i1}) = matrix(CC{i1},CC{i1})\source(CC{i1});
    end
    %
    % COMPUTING STOP CRITERIA
    %
    error = 0;
    for j1=1:nv
        error = error + norm(Raux{j1}-R{j1});
    end
    error = sqrt(error);
    iter = iter + 1;
    if iter==mxit, error = 0.0; end
end
return
```

In turn, we reproduce here the second sub-step of the time integration algorithm proposed by Bathe [15]. Remember that you can use in this framework your favorite time integration scheme (Newmark, HHT, ...)

```
function [R,iter] = enrichment_substep2(K,M,V,num_iter,FV,R,FR,CC,deltat)
% [R,iter] = enrichment_substep2(K,M,V,num_iter,FV,R,FR,CC,deltat)
% Compute the new R, S, functions to enrich the PGD solution for SubStep 2
% in Dynamic problems.
% Universidad de Zaragoza - 2015
iter = 1; TOL = 1.0E-4; error = 1.0; % Inicializating values.
nv = size(FV,1); % # Number of variables for the PGD.
mxit = 11; % # of possible iterations for the fixed point algorithm.
%
```

```
% FIXED POINT ALGORITHM
%
while abs(error)>TOL
    Raux = R; % Updating R(S) last values.
    for i1=1:nv
        %
        % MATRIX COMPUTATION
        %
        matrix1 = 9.0/deltat/deltat; % Constant of mass contribution.
        matrix2 = 1.0; % Constant of stiffness contribution.
        for j1=1:nv
            if j1~=i1
                matrix1 = matrix1.*(R{j1}'*M{j1}*R{j1});
                matrix2 = matrix2.*(R{j1}'*K{j1}*R{j1});
            end
        end
        matrix = M{i1}.*matrix1 + K{i1}.*matrix2;
        %
        % SOURCE COMPUTATION
        %
        source = 0.0;
        for k1-1:size(V{1},2) % Loop over number of functions of the source
            sourceval = 1.0;
            for j1=1:nv
                if j1~=i1
                    sourceval = sourceval.*(R{j1}'*V{j1}(:,k1));
                end
            end
            source = source + sourceval.*V{i1}(:,k1);
        end
        for k1=1:3:(nv-3)
            value1 = 19/deltat/deltat;% Constant value for U_n contribution
            value2 = 5/deltat; % Constant value for V_n contribution.
            value3 = 28/deltat/deltat; % Constant value for U_{n+1/2}.
            for j1=1:nv
                if j1~=i1
                    value1 = value1.*(R{j1}'*M{j1}*FR{j1}(:,k1));
                    value2 = value2.*(R{j1}'*M{j1}*FR{j1}(:,k1+1));
                    value3 = value3.*(R{j1}'*M{j1}*FR{j1}(:,k1+2));
                end
            end
            source = source - value1.*(M{i1}*FR{i1}(:,k1)) - ...
                value2.*(M{i1}*FR{i1}(:,k1+1)) +...
                value3.*(M{i1}*FR{i1}(:,k1+2));
        end
        %
        % CONTRIBUTION TO SOURCE OF KNOWNING SOLUTION
        %
        for i2=1:num_iter-1
            value1 = 9/deltat/deltat; % Constant of mass contribution.
            value2 = 1.0; % Constant of stiffness contribution.
            for j1=1:nv
                if j1~=i1
                    value1 = value1.*(R{j1}'*M{j1}*FV{j1}(:,i2));
                    value2 = value2.*(R{j1}'*K{j1}*FV{j1}(:,i2));
                end
            end
            source = source - (M{i1}*FV{i1}(:,i2)).*value1 - ...
                (K{i1}*FV{i1}(:,i2)).*value2;
        end
        %
        % SOLVE THE R{i1} VARIABLE
        %
        R{i1}(CC{i1}) = matrix(CC{i1},CC{i1})\source(CC{i1});
    end
    %
    % COMPUTING STOP CRITERIA
```

```
    %
    error = 0;
    for j1=1:nv
        error = error + norm(Raux{j1}-R{j1});
    end
    error = sqrt(error);
    iter = iter + 1;
    if iter==mxit, error = 0.0; end
end
return
```

In subroutine `fem3D.m` we accomplish traditional FE computations regarding stiffness matrix, etc., for linear tetrahedrons.

```
function [A,N] = fem3D
% function [A,N] = fem3D
% Finite Element Method for Tetrahedron. The vargout are the Stifness and
% Mass matrices for spacial variables.
% Universidad de Zaragoza - 2015
global E nu coords tet

dof = 3; % Degrees of Freedom per node.
numNodos = size(coords,1); % # of nodes.
numTet = size(tet,1); % # of 3D elements.
A = zeros(dof*numNodos); N = zeros(dof*numNodos); % Allocate memory.
%
% MATERIAL MATRIX
%
D = zeros(6); cte = E*(1-nu)/(1+nu)/(1-2*nu);
D(1) = cte; D(8) = D(1); D(15) = D(1);
D(2) = cte*nu/(1-nu); D(3) = D(2); D(7) = D(2); D(9) = D(2); D(13) = D(2);
D(14) = D(2); D(22) = cte*(1-2*nu)/2/(1-nu); D(29) = D(22); D(36) = D(22);
%
% INTEGRATION POINTS
%
sg = zeros(3,4); wg = 1./24.*ones(4,1); nph = numel(wg);
a = (5.0 - sqrt(5))/20.0; b= (5.0 + 3.0*sqrt(5))/20.0;
sg(:,1) = [a; a; a]; sg(:,2) = [a; a; b];
sg(:,3) = [a; b; a];  sg(:,4) = [b; a; a];
%
% FINITE ELEMENT LOOP
%
for i1=1:numTet
    elnodes = tet(i1,:);
    xcoord = coords(elnodes,:);
    K = zeros(dof*4); KK = zeros(dof*4);
    %
    % JACOBIAN
    %
    v1 = xcoord(1,:)-xcoord(2,:); v2 = xcoord(2,:)-xcoord(3,:);
    v3 = xcoord(3,:)-xcoord(4,:);
    jcob = abs(det([v1;v2;v3]));
    %
    % SHAPE FUNCTION CONSTANTS
    %
    a1 = det([xcoord(2,:); xcoord(3,:); xcoord(4,:)]);
    a2 = -det([xcoord(1,:); xcoord(3,:); xcoord(4,:)]);
    a3 = det([xcoord(1,:); xcoord(2,:); xcoord(4,:)]);
    a4 = -det([xcoord(1,:); xcoord(2,:); xcoord(3,:)]);
    b1 = -det([1 xcoord(2,2:end); 1 xcoord(3,2:end); 1 xcoord(4,2:end)]);
    b2 = det([1 xcoord(1,2:end); 1 xcoord(3,2:end); 1 xcoord(4,2:end)]);
    b3 = -det([1 xcoord(1,2:end); 1 xcoord(2,2:end); 1 xcoord(4,2:end)]);
    b4 = det([1 xcoord(1,2:end); 1 xcoord(2,2:end); 1 xcoord(3,2:end)]);
    c1 = det([1 xcoord(2,1) xcoord(2,end); 1 xcoord(3,1) xcoord(3,end);...
        1 xcoord(4,1) xcoord(4,end)]);
    c2 = -det([1 xcoord(1,1) xcoord(1,end); 1 xcoord(3,1) xcoord(3,end);...
```

```
    1 xcoord(4,1) xcoord(4,end)]);
c3 = det([1 xcoord(1,1) xcoord(1,end); 1 xcoord(2,1) xcoord(2,end);...
    1 xcoord(4,1) xcoord(4,end)]);
c4 = -det([1 xcoord(1,1) xcoord(1,end); 1 xcoord(2,1) xcoord(2,end);...
    1 xcoord(3,1) xcoord(3,end)]);
d1 = -det([1 xcoord(2,1:end-1); 1 xcoord(3,1:end-1);...
    1 xcoord(4,1:end-1)]);
d2 = det([1 xcoord(1,1:end-1); 1 xcoord(3,1:end-1);...
    1 xcoord(4,1:end-1)]);
d3 = -det([1 xcoord(1,1:end-1); 1 xcoord(2,1:end-1);...
    1 xcoord(4,1:end-1)]);
d4 = det([1 xcoord(1,1:end-1); 1 xcoord(2,1:end-1);...
    1 xcoord(3,1:end-1)]);
%
% INTEGRATION POINTS LOOP
%
for j1=1:nph
    chi = sg(3*(j1-1)+1);
    eta = sg(3*(j1-1)+2);
    tau = sg(3*j1);
    %
    % GEOMETRY APPROACH
    %
    SHPa(4) = tau;
    SHPa(3) = eta; SHPa(2) = chi; SHPa(1) = 1.-chi-eta-tau;
    chiG = 0.0; etaG = 0.0; tauG = 0.0;
    for k1=1:4
        chiG = chiG + SHPa(k1)*xcoord(k1,1);
        etaG = etaG + SHPa(k1)*xcoord(k1,2);
        tauG = tauG + SHPa(k1)*xcoord(k1,3);
    end
    %
    % SHAPE FUNCTION COMPUTATION
    %
    SHP(1) = (a1 + b1*chiG + c1*etaG + d1*tauG)/jcob;
    dSHPx(1) = b1/jcob; dSHPy(1) = c1/jcob; dSHPz(1) = d1/jcob;
    SHP(2) = (a2 + b2*chiG + c2*etaG + d2*tauG)/jcob;
    dSHPx(2) = b2/jcob; dSHPy(2) = c2/jcob; dSHPz(2) = d2/jcob;
    SHP(3) = (a3 + b3*chiG + c3*etaG + d3*tauG)/jcob;
    dSHPx(3) = b3/jcob; dSHPy(3) = c3/jcob; dSHPz(3) = d3/jcob;
    SHP(4) = (a4 + b4*chiG + c4*etaG + d4*tauG)/jcob;
    dSHPx(4) = b4/jcob; dSHPy(4) = c4/jcob; dSHPz(4) = d4/jcob;
    %
    % B MATRIX COMPUTATION
    %
    B = [dSHPx(1) 0 0 dSHPx(2) 0 0 dSHPx(3) 0 0 dSHPx(4) 0 0; ...
        0 dSHPy(1) 0 0 dSHPy(2) 0 0 dSHPy(3) 0 0 dSHPy(4) 0;...
        0 0 dSHPz(1) 0 0 dSHPz(2) 0 0 dSHPz(3) 0 0 dSHPz(4);...
        dSHPy(1) dSHPx(1) 0 dSHPy(2) dSHPx(2) 0 dSHPy(3) dSHPx(3)...
        0 dSHPy(4) dSHPx(4) 0;...
        dSHPz(1) 0 dSHPx(1) dSHPz(2) 0 dSHPx(2) dSHPz(3) 0 dSHPx(3)...
        dSHPz(4) 0 dSHPx(4);...
        0 dSHPz(1) dSHPy(1) 0 dSHPz(2) dSHPy(2) 0 dSHPz(3) dSHPy(3)...
        0 dSHPz(4) dSHPy(4)];
    %
    % MASS MATRIX
    %
    M = [SHP(1) 0 0; 0 SHP(1) 0; 0 0 SHP(1); SHP(2) 0 0; 0 SHP(2) 0;...
        0 0 SHP(2); SHP(3) 0 0; 0 SHP(3) 0; 0 0 SHP(3); SHP(4) 0 0;...
        0 SHP(4) 0; 0 0 SHP(4)]';
    K = K + B'*D*B*jcob*wg(j1);  % Element Stiffness matrix.
    KK = KK + M'*M*jcob*wg(j1); % Element Mass matrix.
end
% System degrees of freedom associated with each element.
index = [3*elnodes-2;3*elnodes-1;3*elnodes];
index = reshape(index,1,4*dof);
% Assembling of the system stiffness matrix.
```

```
    A(index,index) = A(index,index) + K;
    N(index,index) = N(index,index) + KK;
end
return
```

Function `evaluate_shpfunc.m` computes finite element one-dimensional shape functions.

```
function S = evaluate_shpfunc(coor,cx)
% function S = evaluate_shpfunc(coor,cx)
% Compute approach value for cx coordinate respect coor points using 1D
% Shape functions
% Universidad de Zaragoza - 2015
tam = numel(coor); % # number of nodes.
TOL = 1.0E-8; % Tolerance.
S = zeros(tam,1); % Allocating memory.
idx = 0; % Index of elemnt where cx is.
for i1=1:numel(coor)-1 % # Elements Loop to find with in
    if coor(i1)-TOL<=cx && coor(i1+1)>=cx
        idx = i1;
        break;
    end
end
% A = find(coor(coor>=cx+TOL));
% idx = A(end); % Idx is element containing cx.
if idx~=tam && idx~=0
    L = coor(idx+1)-coor(idx); % Linear approximation.
    S(idx+1) = (cx-coor(idx))/L;
    S(idx) = (coor(idx+1)-cx)/L;
elseif idx==0 % Not found any element
    S(1) = 1.0;
    disp('It␣is␣possible␣that␣discretization␣is␣not␣enough');
else % Last element
    S(tam) = 1.0;
end
return
```

This same code has been employed, for instance, to generate an interactive web page that makes use very efficiently of the PGD concepts. It can be found at http://amb.unizar.es/beamdyn.htm. It represents, see Fig. 5.4, a linear elastic beam that can be interactively manipulated with the aid of the mouse. It places a vertical, unitary load on the upper surface of the beam. It can be noticed how the very efficient time integration algorithms employed for its construction make it possible to remain vibrating for very long times with minimal numerical dissipation.

Execution of the program produces a window with the tip displacement, see Fig. 5.3. Minimal deviation with respecto to the full-order problem solution is obtained. In any case, higher accuracy can be obtained by employing more POD modes, for instance.

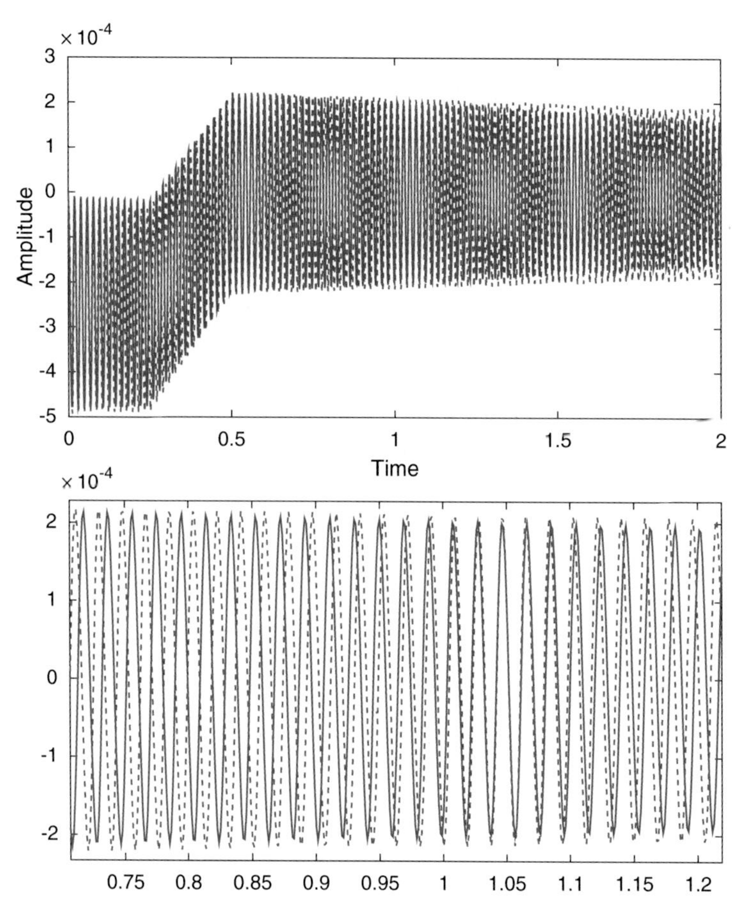

Fig. 5.3 Results of the vibration of a beam (*top*) and detail of the reference solution versus the approximated one (*bottom*)

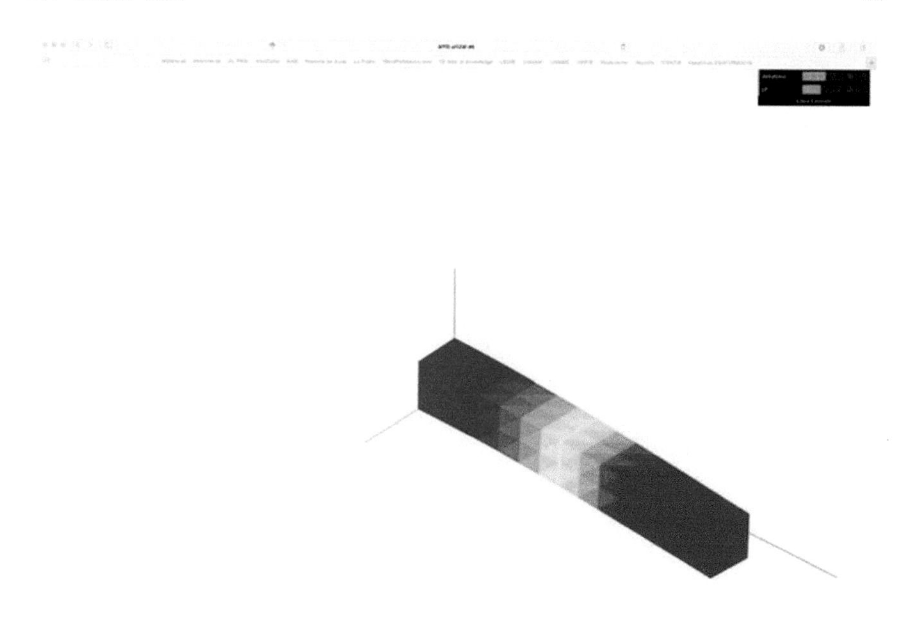

Fig. 5.4 Web implementation of the algorithm here described so as construct an interactive simulator. It represents a linear elastic beam. With the help of the mouse a vertical load is placed on the upper surface of the beam. It can be downloaded from http://amb.unizar.es/beamdyn.htm

References

1. Abichou H, Zahrouni H, Potier-Ferry M (2002) Asymptotic numerical method for problems coupling several nonlinearities. Comput Methods Appl Mech Eng 191(51–52):5795–5810
2. Aguado JV, Huerta A, Chinesta F, Cueto E (2015) Real-time monitoring of thermal processes by reduced-order modeling. Int J Numer Methods Eng 102(5, SI):991–1017
3. Alfaro I, Gonzalez D, Zlotnik S, Diez P, Cueto E, Chinesta F (2015) An error estimator for real-time simulators based on model order reduction. Adv Model Simul Eng Sci 2:30
4. Ammar A, Chinesta F, Diez P, Huerta A (2010) An error estimator for separated representations of highly multidimensional models. Comput Methods Appl Mech Eng 199(25–28):1872–1880
5. Ammar A, Cueto E, Chinesta F (2012) Reduction of the chemical master equation for gene regulatory networks using proper generalized decompositions. Int J Numer Methods Biomed Eng 28(9):960–973
6. Ammar A, Mokdad B, Chinesta F, Keunings R (2006) A new family of solvers for some classes of multidimensional partial differential equations encountered in kinetic theory modeling of complex fluids. J Non-Newton Fluid Mech 139:153–176
7. Ammar A, Mokdad B, Chinesta F, Keunings R (2007) A new family of solvers for some classes of multidimensional partial differential equations encountered in kinetic theory modeling of complex fluids. Part II. Transient simulation using space-time separated representations. J. Non-Newton Fluid Mech 144:98–121
8. Ammar A, Chinesta F, Cueto E (2011) Coupling finite elements and proper generalized decompositions. Int J Multiscale Comput Eng 9(1):17–33
9. Ammar A, Chinesta F, Cueto E, Doblare M (2012) Proper generalized decomposition of time-multiscale models. Int J Numer Methods Eng 90(5):569–596
10. Ammar A, Huerta A, Chinesta F, Cueto E, Leygue A (2014) Parametric solutions involving geometry: a step towards efficient shape optimization. Comput Methods Appl Mech Eng 268:178–193
11. Ammar A, Pruliere E, Ferec J, Chinesta F, Cueto E (2009) Coupling finite elements and reduced approximation bases. Eur J Comput Mech 18(5–6):445–463
12. Amsallem D, Farhat Ch (2008) An interpolation method for adapting reduced-order models and application to aeroelasticity. AIAA J 46:1803–1813
13. Barbarulo A, Ladeveze P, Riou H, Kovalevsky L (2014) Proper generalized decomposition applied to linear acoustic: a new tool for broad band calculation. J Sound Vib 333(11):2422–2431
14. Barrault M, Maday Y, Nguyen NC, Patera AT (2004) An 'empirical interpolation' method: application to efficient reduced-basis discretization of partial differential equations. C R Math 339(9):667–672
15. Bathe KJ (2007) Conserving energy and momentum in nonlinear dynamics: a simple implicit time integration scheme. Comput Struct 85:437–445

E. Cueto et al., *Proper Generalized Decompositions*, SpringerBriefs
in Applied Sciences and Technology, DOI 10.1007/978-3-319-29994-5

16. Bognet B, Bordeu F, Chinesta F, Leygue A, Poitou A (2012) Advanced simulation of models defined in plate geometries: 3d solutions with 2d computational complexity. Comput Methods Appl Mech Eng 201–204:1–12
17. Bonet J, Wood RD (2008) Nonlinear continuum mechanics for finite element analysis. Cambridge University Press
18. Boucinha L, Gravouil A, Ammar A (2013) Space-time proper generalized decompositions for the resolution of transient elastodynamic models. Comput Methods Appl Mech Eng 255:67–88
19. Bouclier R, Louf F, Chamoin L (2013) Real-time validation of mechanical models coupling PGD and constitutive relation error. Comput Mech 52(4):861–883
20. Cao H-L, Potier-Ferry M (1999) An improved iterative method for large strain viscoplastic problems. Int J Numer Methods Eng 44:155–176
21. Chaturantabut S, Sorensen DC (2010) Nonlinear model reduction via discrete empirical interpolation. SIAM J Sci Comput 32:2737–2764
22. Chinesta F, Abisset-Chavanne E, Ammar A, Cueto E (2015) Efficient stabilization of advection terms involved in separated representations of boltzmann and fokker-planck equations. Commun Comput Phys 17(4):975–1006
23. Chinesta F, Ammar A, Cueto E (2010) Proper generalized decomposition of multiscale models. Int J Numer Methods Eng 83(8–9, SI):1114–1132
24. Chinesta F, Ammar A, Cueto E (2010) Recent advances in the use of the proper generalized decomposition for solving multidimensional models. Arch Comput Methods Eng 17(4):327–350
25. Chinesta F, Ammar A, Joyot P (2008) The nanometric and micrometric scales of the structure and mechanics of materials revisited: an introduction to the challenges of fully deterministic numerical descriptions. Int J Multiscale Comput Eng 6:191–213
26. Chinesta F, Leygue A, Bordeu F, Aguado JV, Cueto E, Gonzalez D, Alfaro I, Ammar A, Huerta A (2013) PGD-based computational vademecum for efficient design, optimization and control. Arch Comput Methods Eng 20(1):31–59
27. Chinesta Francisco, Cueto Elias (2014) PGD-based modeling of materials, structures and processes. Springer, Switzerland
28. Chinesta F, Keunings R, Leygue A (2014) The proper generalized decomposition for advanced numerical simulations. Springer, Switzerland
29. Chinesta F, Ladeveze P, Cueto E (2011) A short review on model order reduction based on proper generalized decomposition. Arch Comput Methods Eng 18:395–404
30. Chinesta F, Magnin M, Roux O, Ammar A, Cueto E (2015) Kinetic theory modeling and efficient numerical simulation of gene regulatory networks based on qualitative descriptions. Entropy 17(4):1896–1915
31. Cochelin B, Damil N, Potier-Ferry M (1994) The asymptotic numerical method: an efficient perturbation technique for nonlinear structural mechanics. Rev Eur Elem Finis 3:281–297
32. Cremonesi M, Neron D, Guidault PA, Ladeveze P (2013) A PGD-based homogenization technique for the resolution of nonlinear multiscale problems. Comput Methods Appl Mech Eng 267:275–292
33. Cueto E, Chinesta F (2014) Real time simulation for computational surgery: a review. Adv Model Simul Eng Sci 1(1):11
34. Pearson K (1901) F.R.S. Liii. on lines and planes of closest fit to systems of points in space. Philos Mag Ser 6, 2(11):559–572
35. Ghnatios Ch, Masson F, Huerta A, Leygue A, Cueto E, Chinesta F (2012) Proper generalized decomposition based dynamic data-driven control of thermal processes. Comput Methods Appl Mech Eng 213–216:29–41
36. Ghnatios Ch, Masson F, Huerta A, Leygue A, Cueto E, Chinesta F (2012) Proper generalized decomposition based dynamic data-driven control of thermal processes. Comput Methods Appl Mech Eng 213–216:29–41
37. Giacoma A, Dureisseix D, Gravouil A, Rochette M (2015) Toward an optimal a priori reduced basis strategy for frictional contact problems with LATIN solver. Comput Methods Appl Mech Eng 283:1357–1381

38. Gonzalez D, Ammar A, Chinesta F, Cueto E (2010) Recent advances on the use of separated representations. Int J Numer Methods Eng 81(5):
39. Gonzalez D, Masson F, Poulhaon F, Cueto E, Chinesta F (2012) Proper generalized decomposition based dynamic data driven inverse identification. Math Comput Simul 82:1677–1695
40. Gonzalez D, Alfaro I, Quesada C, Cueto E, Chinesta F (2015) Computational vademecums for the real-time simulation of haptic collision between nonlinear solids. Comput Methods Appl Mech Eng 283:210–223
41. Gonzalez D, Cueto E, Chinesta F (2014) Real-time direct integration of reduced solid dynamics equations. Int J Numer Methods Eng 99(9):633–653
42. Henneron T, Clenet S (2015) Proper generalized decomposition method applied to solve 3-d magnetoquasi-static field problems coupling with external electric circuits. IEEE Trans Magn 51(6):
43. Karhunen K (1946) Uber lineare methoden in der wahrscheinlichkeitsrechnung. Ann Acad Sci Fennicae Al Math Phys 37
44. Ladeveze P (1999) Nonlinear computational structural mechanics. Springer, New York
45. Ladeveze P, Passieux J-C, Neron D (2010) The latin multiscale computational method and the proper generalized decomposition. Comput Methods Appl Mech Eng 199(21–22):1287–1296
46. Ladeveze P, Chamoin L (2011) On the verification of model reduction methods based on the proper generalized decomposition. Comput Methods Appl Mech Eng 200(23–24):2032–2047
47. Le Bris C, Lelièvre T, Maday Y (2009) Results and questions on a nonlinear approximation approach for solving high-dimensional partial differential equations. Constr Approximation 30:621–651. doi:10.1007/s00365-009-9071-1
48. Loève MM (1963) Probability theory. The university series in higher mathematics, 3rd edn. Van Nostrand, Princeton
49. Lorenz EN (1956) Empirical orthogonal functions and statistical weather prediction. MIT, Departement of Meteorology, Scientific Report Number 1, Statistical Forecasting Project
50. Ly HV, Tran HT (2005) Modeling and control of physical processes using proper orthogonal decomposition. Math Comput Model 33:223–236
51. Modesto D, Zlotnik S, Huerta A (2015) Proper generalized decomposition for parameterized helmholtz problems in heterogeneous and unbounded domains: application to harbor agitation. Comput Methods Appl Mech Eng 295:127–149
52. Moitinho de Almeida JP (2013) A basis for bounding the errors of proper generalised decomposition solutions in solid mechanics. Int J Numer Methods Eng 94(10):961–984
53. Nguyen NC, Patera AT, Peraire J (2008) A 'best points' interpolation method for efficient approximation of parametrized functions. Int J Numer Methods Eng 73(4):521–543
54. Niroomandi S, Alfaro I, Cueto E, Chinesta F (2008) Real-time deformable models of non-linear tissues by model reduction techniques. Comput Methods Programs Biomed 91(3):223–231
55. Niroomandi S, Alfaro I, Gonzalez D, Cueto E, Chinesta F (2012) Real-time simulation of surgery by reduced-order modeling and x-fem techniques. Int J Numer Methods Biomed Eng 28(5):574–588
56. Niroomandi S, González D, Alfaro I, Bordeu F, Leygue A, Cueto E, Chinesta F (2013) Real-time simulation of biological soft tissues: a PGD approach. Int J Numer Methods Biomed Eng 29(5):586–600
57. Niroomandi S, Gonzalez D, Alfaro I, Cueto E, Chinesta F (2013) Model order reduction in hyperelasticity: a proper generalized decomposition approach. Int J Numer Methods Eng 96(3):129–149
58. Niroomandi Siamak, Alfaro Iciar, Cueto Elias, Chinesta Francisco (2010) Model order reduction for hyperelastic materials. Int J Numer Methods Eng 81(9):1180–1206
59. Quesada C, Gonzalez D, Alfaro I, Cueto E, Chinesta F (2015) Computational vademecums for real-time simulation of surgical cutting in haptic environments. Int J Numer Methods Eng (Submitted)
60. Quesada C, González D, Alfaro I, Cueto E, Huerta A, Chinesta F (2015) Real-time simulation techniques for augmented learning in science and engineering. Vis Comput 1–15

61. Vitse M, Neron D, Boucard P-A (2014) Virtual charts of solutions for parametrized nonlinear equations. Comput Mech 54(6):1529–1539
62. Yvonnet J, Zahrouni H, Potier-Ferry M (2007) A model reduction method for the post-buckling analysis of cellular microstructures. Comput Methods Appl Mech Eng 197:265–280
63. Zlotnik S, Diez P, Modesto D, Huerta A (2015) Proper generalized decomposition of a geometrically parametrized heat problem with geophysical applications. Int J Numer Methods Eng 103(10):737–758

Index

E. Cueto et al., *Proper Generalized Decompositions*, SpringerBriefs in Applied Sciences and Technology, DOI 10.1007/978-3-319-29994-5